OUT OF THE CRATER

Mount Rainier, Washington—massive and beautiful. At its foot I have rested along the trails in the silence of the forest and watched eagles soar. I have watched cliffs of ice at glaciers' ends crackling and then crashing to the valley floor and have crossed creeks on stones rounded by rushing debris floods. I have watched its white slopes painted red by the rising sun and have sat by my campfire in the night as the moon's ghostly light reflected from its snowy summit. (Photo: Richard V. Fisher.)

Richard V. Fisher

Out of the Crater

Chronicles of a Volcanologist

Princeton University Press, Princeton and Oxford

Copyright © 1999 by Princeton University Press
Published by Princeton University Press, 41 William Street,
Princeton, New Jersey 08540
In the United Kingdom: Princeton University Press,
3 Market Place, Woodstock, Oxfordshire OX20 1SY

Fourth printing, and first paperback printing, 2000

Paperback ISBN 0-691-07017-2

*The Library of Congress has cataloged the cloth edition
of this book as follows*
Fisher, Richard V. (Richard Virgil), 1928–
Out of the crater : chronicles of a volcanologist /
Richard V. Fisher
p.   cm.
Includes index.
ISBN 0-691-00226-6 (cloth : alk. paper).
1. Fisher, Richard V. (Richard Virgil), 1928–
2. Volcanologists—United States—Biography.   3. Volcanoes.
I. Title.
QE22.F44A3   1999
551.2'1'092—dc21   98-17838
[B]

This book has been composed in Palatino

The paper used in this publication meets the minimum requirements
of ANSI/NISO Z39.48-1992 (R1997) (*Permanence of Paper*)

www.pup.princeton.edu

Printed in the United States of America

10   9   8   7   6   5   4

◇◆◇◆◇◆◇◆◇◆◇◆◇◆◇◆◇◆◇◆◇◆◇◆◇◆◇◆◇◆◇◆◇◆◇

This book is dedicated to
Beverly, my friend, companion, and wife.
Geological field studies separated us in
the early summers of our life together,
but after our children were grown,
she accompanied me to China, England,
France, Germany, Hawaii, Italy,
Martinique, Oregon, St. Vincent, Spain,
and Washington and opened my eyes
to the poetry and wonder of the art and
history of those volcanic lands.

Contents

Acknowledgments

Until I wrote this book, my training and experience in writing, was professional science writing—clipped, trimmed, hard-edged, and data driven. I had no experience with what I call "recreational writing," and I was apprehensive until my wife Beverly—honest to a fault and a hard critic when it comes to judging my work—put my mind at ease. She liked it! That gave me the fire and confidence to continue toward my goal of publishing a short personal account of my career. Marilyn Power Scott, friend, critic, and editor with red pen, went through all of my words, scattering them about like ten pins, forcing me to look hard at *how* I said what I meant to say. After Beverly and Marilyn gave their blessings, I then searched for a publisher—not a small task—and was rescued by Jack Repcheck, senior editor for earth science at Princeton University Press, who asked to see the manuscript (and later critically edited it). He liked it and sent it to two reviewers, Dr. Peter Schiffman of the University of California at Davis and Dr. Barton S. Martin of Ohio Wesleyan University, each of whom recommended publication. Hans-Ulrich Schmincke, my colleague-in-research in Germany, kindly helped with the chapter on volcanoes in Europe. My friend, colleague, and former student Michael Ort, professor at Northern Arizona State University, reviewed the chapter on calderas to jog my failing memory of our high-Andes trip to Cerro Panizos, Argentina. Another friend, colleague, and former student, Grant Heiken, read the text to check for obvious errors of fact. I am indebted to these spirits of good will and professionalism.

OUT OF THE CRATER

◇◇◇◇◇◇◇◇◇◇◇◇◇◇◇◇◇◇◇◇◇◇◇◇◇◇◇◇◇◇◇◇◇◇◇

Prologue

At the age of twenty, with a great deal of luck, I stumbled upon the profession of geology, and for nearly fifty years I have courted the earth, roaming its deserts and forests, its mountains and volcanoes. I was often alone with the elements, extracting stories from the rocks, making new discoveries and building on those of my contemporaries and predecessors. It has been the most enjoyable of occupations, solving unsolved mysteries of the earth and thinking of new ideas—the lifeblood of science.

Many facets of geology have seeped into my subconsciousness and have become second nature to me. Like other geologists on family vacations, I am able to tell story after story of how certain features of the visible landscape were formed—in glaciated terrain, lake basins, mountain peaks, desert terrain, volcanic lands, river lowlands, and broad and narrow valleys. Some people say that knowing the scientific origin of natural features spoils the spirituality and enjoyment of the earth, but this is not so for geologists.

One of my greatest lessons was learning to appreciate geologic time, which is the key to understanding geologic processes. Given the immensity of time, great changes occur within and upon the earth by processes so slow that they are unnoticed in the lifetime of a person. Human beings can barely, if at all, detect normal changes of the earth—for example, momentous "upheavals" of the Himalyan Mountains—as they occur. A rate of uplift on the order of a tenth of an inch per year can raise mountains to over eight thousand feet in one million years—slow to us, but very fast on the possibly infinite time scale of the universe. Our lives are single frames in the moving picture of the constantly changing earth. The Grand Canyon was eroded, granule by granule, by the Colorado River, with erosion accelerating as steep rocky cliffs, carved by water, slid or avalanched into the deep chasm. The water of the powerful, relentless Colorado carried away the rocks, sand, and mud southward to construct a delta at the Gulf of California.

Volcanoes, however, are geological systems operating in real time. You can see them create new rocks on the earth's surface and, in areas such as Hawaii, new land in the oceans. In other places, erosion has stripped volcanoes down to their roots, exposing ancient volcanic necks where the surrounding land has been worn away. Understanding volcanoes—both new and old—and, ultimately, volcanism requires that, for each volcano, volcanologists determine the chemistry, unravel the history, and discover the eruptive cycles and the types of eruptions that have occurred.

For most of my professional career, I have worked in remote lands, including the serenity of Douglas-fir–covered canyons and the ridges of the Cascade Mountains in the shadow of Mount Rainier; the sagebrush-, juniper-, and pine-covered mountains of central Oregon; the southern desert coast of the island of Hawaii; the tropical island of Martinique; and the false desert made by the 1980 eruption of Mount St. Helens.

Gaia

To the Greeks, our home planet was *Gaia* (or Gaea), goddess of the earth, from which the name *geology* was derived. The word *Earth* came later from the Middle English. British atmospheric scientist James Lovelock and American microbiologist Lynn Margulis formulated the Gaia Hypothesis, which states that the earth's surface environment is controlled by plants, animals, and microorganisms. Although this reciprocal arrangement between the earth's environment and living things may be true, it does not necessarily follow that the earth is an animate object similar to organic life, as some Gaia followers believe. The earth regulates itself by a complex system of corrective chemical mechanisms that adjust to changes in temperature and chemistry of the environment (salinity, acidity, etc.), as do living organisms, but the earth cannot reproduce itself. Many Gaian followers believe that if the planet functions like an integrated body, the earth has vulnerable places equivalent to vital organs, where serious harm may spread to the whole body. Tropical rain forests, for example, are believed vital for the life of the earth, like a liver that is necessary for life in an animal. If such a vital organ is destroyed, the entire organism can die or be crippled.

Some advocates have carried the Gaian hypothesis of the earth's ultra-thin-skinned biological balance into the realm of earth goddess worship, claiming that Gaia is not only alive, she is conscious. Although this claim is scientifically invalid, it is a belief system that has value because Gaian followers advocate preserving and protecting earth systems, and therefore join with global environmentalists, geologists, other natural scientists, and many others in that endeavor. Geologists have an inside track to the understanding of Gaia. Many, if not all, geologists will ridicule the humanization of the earth, but they cannot deny that life evolved from the earth's inorganic surface processes and that life continues to take energy from it.

Background

Let me provide some background for the rest of this book by introducing the volcanic rocks and the types of volcanoes that have captivated me throughout my career. These are the products of two main types of eruptions. The first type is an eruption that occurs when *magma* (see Glossary for italicized terms) encounters and mixes with water in the environment, such as a lake or an ocean; a *hydrovolcanic eruption*. The second type is a highly explosive eruption caused by the sudden expansion of internal gases from within magma; a *magmatic eruption*. Hydrovolcanic eruptions produce wide-diameter craters from shallow explosions. Magmatic eruptions produce many kinds of volcanoes and craters. Both kinds of eruptions can produce horrendously powerful volcanic hurricanes known as *pyroclastic flows and surges* (explained later), subjects that have comprised most of my professional research activity.

This book reveals a few of my adventures, scientific and cultural, related to the study of volcanoes and volcanic rocks—adventures that span fifty years. This uncommonly pleasurable occupation may give the impression that my research grant money was spent on vacations, but that is an artifact of the writing, which leans heavily on remembered adventures—scientific and cultural—rather than on the daily work of gathering detailed data in the field.

The purpose of my research is to discover the processes of emplacement of pyroclastic flows and surges, which are useful for helping people survive volcanic disasters. This rises out of a curiosity about how things work, and I treasure the great pleasure of discovery, the adventure of travel, and intimacy with earth's quiet and blessed beauty. It gives me considerable satisfaction to know that some results of my curiosity have led to practical use and that other geologists have built on them in their own research.

This book is mostly about walks and talks with some of the volcanoes that I have studied, visited, climbed, or contemplated. It is not a scientific document, for I give my own points of view without proofs and rarely discuss contrary ideas (which are found in other books and journals). This, too, is a partial autobiography to explain to my children what I was doing those too many summers absent from home and to let my grandchildren and their children know something about their grandfather.

Following most of the chapters are lists of some recommended books. These suggested readings do not specifically discuss the content of the chapters, but parts of the readings include subjects within the chapters. At the end of the book is a glossary of geologic terms (words italicized in text at first usage or where they are defined in text) for quick reference.

Suggested Reading

Bullard, F. M. *Volcanoes.* Austin: University of Texas Press, 1968.

Lovelock, J. *The Ages of Gaia: A Biography of Our Living Earth.* New York: W. W. Norton, 1988.

Chapter One

◇◇◇◇◇◇◇◇◇◇◇◇◇◇◇◇◇◇◇◇◇◇◇◇◇◇◇◇◇◇◇◇◇◇

Los Alamos, New Mexico

It was my fiftieth birthday, a warm August 8, 1978. That morning I had decided to do field work on San Ildefonso Pueblo land in New Mexico, without authorization from the San Ildefonso Pueblo tribal council. I made the decision with considerable anxiety, for I had heard stories of people who had entered the property without permission only to have their cameras, wallets, and other private possessions taken from them. My camera was indispensable, and my field notebook was filled with geological data—there was no other copy. I was tracing a *volcanic ash* layer that was exposed in gulleys and canyon walls across the Pajarito Plateau, near the towns of White Rock and Los Alamos, much of which is owned by the San Ildefonso Pueblo. Gambling that I would not be discovered, I decided to explore that part of the plateau anyway. I was less sensitive in those days, and apologize for my trespass, for I know now that sacred lands need to be honored.

I walked across an open, gullied, upland surface with low shrubs, feeling very noticeable, expecting at any minute to be surrounded by hostile elders, but the land was empty of people, and I saw only a few rabbits. I walked about half a mile across the flatland and then entered a small canyon and started up the other side. Halfway up, I heard voices and the bark of a dog. They could only be from people from the pueblo, so I crouched behind a very small bush and froze. Two horsemen herding three cows came over the crest of a hill on the other side of the canyon at eye level with me, but they were deep in conversation. Had they flicked their eyes upward, they would have seen me crouching like a giant ball behind the tiny bush, but, preoccupied, they rode down the opposite ridge below my position without looking up. As they crossed my path, the dog stopped, vigorously sniffed at my trail, and began to follow my smell up the slope. I closed my eyes, knowing that I was about to be discovered. But one of the riders whistled, and, to my great relief, the dog turned and ran after the men.

The area of the plateau that I intended to explore that day was a long finger of rolling flatland about three miles long and a few hundred feet wide, bounded by canyons with steep cliffs, three to four hundred feet high, carved into the plateau by small tributaries of the Rio Grande, the ribbon of water that snakes from Colorado to the Gulf of Mexico. The plateau, covered with junipers, chaparral, and piñons, is poor grazing land for cattle and therefore little used by the people of the Pueblo. My gamble paid off, and I did not meet another human being that day but instead, discovered their ancestors, ghosts of the ancient Anasazi people who lived in the region from about A.D. 1 to A.D. 1300. I walked in their foot paths, worn six to twelve inches deep in the soft *tuff* by bare and sandal-clad feet (fig. 1). Walking the trails in the hot sun and gentle breezes brought a closeness of spirit with the men, women, and children of that long-ago world. I encountered many shallow caves in the tuff along the base of the cliffs, former homes of the Anasazi. Some of the caves had blackened ceilings. Were they blackened by Anasazi fires or fires from modern hunters? The shape of some caves had been modified for greater comfort—surely by the Anasazi.

To celebrate my birthday, I spent a relaxed lunch in one of the caves, dozing lazily and gazing at the landscape, imagining the sounds of the children at play and watching the descendants of birds that had entertained them with their songs so many hundreds of years ago. I imagined the laughter and the joys, the

Figure 1. Ancient Anasazi foot path near Los Alamos, New Mexico.
I walked the trails made before A.D. 1300 by generations of Anasazi
people living their daily lives. The trail was worn deep into hardened
volcanic ash by sandaled or bare feet. The trail tells of close communication and interaction among the local population, whose dwellings
were carved by wind and water within cliffs of the same rock type,
and it connected the community with the canyon bottoms where corn
was grown. (Photo: Richard V. Fisher.)

shouting and the sorrows, the arguments that swelled now and again. The people played and loved and knew tragedies in their lives on earth. I mused that they had taken their turns at life on this planet from babyhood, to adults, to elders, and finally back to dust, and now it was my turn.

The entire summer of fieldwork in the open pine forests of the plateaus and canyons around Los Alamos was a delight. Each day in the field was a meditation, reminding me why I had chosen to follow a geological career: to discover relationships of the earth that had been revealed to no one else before.

I was doing research at Los Alamos because of my interest in pyroclastic flows, which have been the basis of my career and will be discussed throughout the book. Let me give a little scientific background here: *Pyroclastic* refers to materials that form by explosive processes of volcanoes, one of the main types being volcanic ash, consisting of small pieces of glass, crystals, and rock fragments blown to bits by volcanic explosions. *Pyroclastic flows* are extremely dangerous, searingly hot, heavier-than-air, hurricanelike density currents made up of abundant particles mixed with volcanic gases exploded from a volcano. These flows are commonly denser than water and flow down valleys. They can travel faster than an atmospheric hurricane over distances greater than a hundred miles. *Pyroclastic surges* are less dense than pyroclastic flows because they contain fewer particles, but they are just as fast and hot, can move turbulently across water or land, and are not confined to valleys, even flowing across ridges. *Ignimbrite* is a layer of ash and larger fragments deposited by a pyroclastic flow. Quite often, the source of ignimbrite of great volume is a giant crater known as a *caldera* (see chapt. 11).

Each outcrop of ignimbrite that I explored at Los Alamos represented another piece of the puzzle that was contributing to the hypothesis beginning to form in my mind. At the end of the summer, the layer that I followed and described led to a revelation about one facet of the mechanism of emplacement of ignimbrite that had not been clearly stated before. The revelation was this: During the movement of some pyroclastic flows, expanding gases containing fine-to-coarse ash blow out from the top of the pyroclastic flow as it moves. The ash gets caught in the volcanic winds that blow it across the newly formed deposits to construct shapes that look like small sand dunes formed by ordinary winds.

At age fifty, I was in the middle of my career as a geologist, with several research accomplishments under my belt, and many new projects waiting ahead. It was ironic that my career had started essentially in this same place—Los Alamos—over thirty years earlier.

Suggested Reading

Bandelier, A. F. *The Delight Makers*. New York: Dodd, Mead, 1918.

Chapter Two

◇◇◇◇◇◇◇◇◇◇◇◇◇◇◇◇◇◇◇◇◇◇◇◇◇◇◇◇◇◇◇◇

Bikini and the Atom Bomb

With great shouting and cursing, twenty young U.S. Army recruits and I were verbally lashed into scrubbing the wooden floor of the barracks while the sergeant squirted water on the floor from a hose, commanding us to sweep it uphill across the concrete floor of the latrine. "Get your asses in gear. Dammit!! Move! Move!! *Move!!!!*" This was our second day at Fort MacArthur Induction Center, San Pedro Harbor, California, and they tried to keep us busy while waiting for assignment to a basic training camp. I was seventeen, and it was January 21, 1946.

"Fisher!! Get down on your goddam knees and push the water with your hands!!"

That day went from bad to worse as I plotted how to get out of my next work detail. Hiding from work alternated with getting found and assigned other work duties—cleaning barracks, picking up litter from the ground, sweeping the streets, and whatever else that could be found. I was issued army clothes, was vaccinated, and learned how to make a bed exactly the right way. The corners were tucked in just right, and a coin could bounce from the tight surface at the center of the blanket.

The induction center was also a mustering-out center for veterans just returning from the South Pacific. The returning veterans of World War II did not have work assignments, so I hung around their barracks trying to look like a war veteran—not easy at seventeen. I got plenty of advice from them about how to avoid work assignments. "Fisher, never volunteer for anything," was the single best piece of advice I received.

The harassment kept on for two weeks before any hope of rescue came my way. One morning after the 5 o'clock roll call, fifty of the newest recruits were called, including me and two young men who had joined with me. We were ordered to headquarters, where we were told to stand in line and wait to be interviewed.

"What for?" I asked.

"Shut up, Fisher," replied the sergeant. "No talking. You'll find out soon enough."

My turn finally came to enter a small room sparsely furnished with a framed picture of President Truman on the wall, a desk, and a lieutenant sitting stiffly behind it with a list on the table and a pencil in his hand. I saluted as best I could for a new recruit. I took an instant dislike to the officer.

"What's your name and serial number, private."

"Fisher, sir. One, nine, two, four, double 'oh', seven, eight."

The lieutenant said, "Would you like to be an MP or a decontaminationist?

"What are those?" I asked.

"Sir," he said.

"Sir! Sorry, sir."

"MP stands for Military Police. A decontaminationist checks for radioactivity in laboratories. Both jobs are at Los Alamos, New Mexico, where they make atom bombs. Which do you want to volunteer for?"

I remembered the advice about volunteering.

"Neither, sir," I said.

"I said, *choose one!*"

The ominous tone and the body language were unmistakably hostile. Another bit of advice I had learned from the returning veterans was, "Never argue with an officer."

I absolutely did *not* want to be a policeman, but did not want to tell him what I didn't want. I had only one choice, and I lied, for I did not want to be a chemist either. "Sir, when I get discharged from the army, I want to go to college and be a chemist. I would like to work in the laboratories." I became a decontaminationist.

Without any basic training, I was shipped within a week to Los Alamos (where I would be doing volcano research on ignimbrite thirty-two years later.) I received thirty minutes of instruction on using a Geiger counter and was introduced to my workplace, a group of prefabricated buildings constructed in haste during World War II. The site was an isolated plateau, where scientists perfected an atom bomb intended for use against the Germans. Instead, the bomb was used to kill thousands of Japanese civilians and to hasten the end of the war with the Japanese military forces. My main duty was to check for radioactive areas on laboratory table tops where chemical research was conducted. Using a grease pencil, I would mark the "hot spots" directly on the chemistry lab tables and enter the number of rads per minute onto a form submitted to lab officials at the end of each day. The nuclear chemists and physicists were required to clean the hot spots.

"God, this is boring," I said. Gene Fox, one of my compatriots from my hometown, Whittier, California, agreed with me. The job was repetitive and did not vary from day to day. I did enjoy the weekends, however, as my friends and I explored the natural caves in the volcanic rocks searching for Indian artifacts, or hiked through the pine forests for the pure joy of exercise and explored the valleys around Los Alamos. Yet I became increasingly restless and bored. I asked for reassignment to Germany and was denied. After three months of grinding boredom, I had an opportunity to become involved with the impending atom bomb tests at Bikini, a coral island atoll just north of the equator. I *volunteered* for housekeeping and maintenance troops, even though I abhorred KP. As it turned out, I was extremely fortunate not to have been assigned as a decontaminationist. Boarding and monitoring radioactive ships at Bikini turned out to be very dangerous, for the radioactivity permeated every pore of the target ships.

One of my good friends, Caldwell Jones, signed up with me, and we left Los Alamos for San Francisco on May 18, 1946. By twice breaking the cardinal military edict of "never volunteer," I had serendipitously chosen a path that was to influence my future research career in geology, before I had even discovered geology! On May 24, we boarded the USS *Haven*, a refitted hospital ship from World War II, and I was assigned to work in the supply room together with another friend from Los Alamos, Wilbur Newton. We issued clothing for men who were to monitor radioactive ships and goggles to scientists and military officers who would watch the atom bomb explosion at Bikini Atoll. We left San Francisco on May 29, arriving in Honolulu on June 5. We left for Bikini Atoll June 6 and arrived June 12.

At 9:30 A.M., we steamed into the lagoon of Bikini Atoll. I stood at the railing of the ship and watched in amazement as we passed dozens of expendable ships of the U.S. Navy and some captured German and Japanese ships. The target ships included battleships, cruisers, destroyers, aircraft carriers, submarines, and many smaller landing craft. The Army Air Force intended to sink them with bombs; the Navy expected the Air Force to fail.

We were busy right up to the day of the first test, but we still found time to swim a little and play baseball. I collected exotic seashells along the lagoon's beach, and explored the island of Bikini. The highest elevation was fourteen feet, and it took me about twenty minutes to cross the island. My walk wound through a coconut palm forest starting at the calm waters of the lagoon and ending on the seaward side of the island, where huge breakers from rough seas crashed incessantly onto the banks of the coral reef about a half mile offshore.

On Monday, July 1, "Able Day," the bomb was detonated shortly after 9 A.M. The USS *Haven* was one of the closest observation ships, but still well over the horizon—we could not see Bikini from our position. Enlisted men were unable to watch the bomb blast because they were not issued the goggles needed to darken the intense light of the bomb, but since I worked in the

supply room, I had issued goggles to myself. In anticipation—
and as a precaution—I closed one eye in case I was blinded by
the blast (I have often wondered how many other men watched
the atom bomb explosion with one eye shut!). I could barely see
the sun through the goggles, but the fireball created by the burst-
ing atomic bomb was much brighter than the sun for a few seconds. The bomb was detonated at 518 feet above the lagoon. The
brilliant hemisphere of light quickly went dark, and I tore the
goggles off to watch the explosion plume make a breathtaking
ascent into the calm blue morning sky. The plume was copper-
colored, and turbulent convolutions were darkly outlined as
they enfolded upon one another. At about forty thousand feet,
moisture condensed over the top of the mushroom cloud and
made a smooth white surface, like the topping of an ice cream
cone. The event unfolded like a slow-motion dream.

The first wave of nontarget fleet vessels reentered Bikini lagoon within two hours to check radioactivity levels of the water,
ships, and islands. Five hours after detonation, the rest of the
fleet was allowed to enter. A brief announcement on the inter-
com said that radiological activity was about as expected, but
no details were given.

As we entered the lagoon, we sailed past the aircraft carrier,
Independence, which was on fire. The stern of the submarine *Skate*
was crushed. Several ships were burning, and I could see considerable superficial damage on others. The battleship *Nevada*
(painted orange) at the center of the fleet was the target ship,
but the bomb inexplicably came down and exploded one-half
mile southwest of the *Nevada*, almost directly above the *Independence*. Only one side of the *Nevada* was blackened by the intense
heat of the fireball. Incredibly, the bomb had sunk only two of
the ninety-five target vessels.

Activity on the *Haven* was at a high point as radiological monitors prepared to leave for various ships of the target fleet. Radioactive monitoring of target vessels began by midafternoon of
that day. Although the project leaders had assured us that most
of the radioactivity had dissipated, I learned that evening that

monitoring the ships was extremely dangerous. Harper, one of the radiation monitors, slept in the bunk next to mine. That evening he told me that the *Prinz Eugen*, a German battleship to which he had been assigned, was very radioactive, and he was certain that he had received a high dosage during his day aboard the ship.

Within two days after the detonation, the waters of the lagoon, most of the ships, and the beaches were pronounced radioactively safe. Having spent four months monitoring radioactive labs at Los Alamos and attending lectures about the effects of radioactivity on the human body, I wondered how the area could be safe after the detonation of a highly radioactive atom bomb, but I accepted the judgment of the men in authority. As a consequence, from July 3 to July 25, I spent seven of my free days swimming in the waters of the lagoon and lying on the beach near the "Coral Reef Tavern" on Bikini along with dozens of other young men.

Thursday, July 25 was "Baker Day." The second atom bomb was set off at 8:35 A.M. Goggles were not needed because the bomb was detonated beneath water. It was put in a concrete container and suspended ninety feet beneath a small landing ship. According to my diary, scientists expected a column of water shaped like a "redwood tree" to rise from the lagoon and shoot up to 1,800 feet, and a wave to sweep over the Island. Our ship was to be eight miles from the target, and all the ships within the lagoon would be in sight of the observers aboard the ships outside of the lagoon.

The events of the blast happened so fast that they are difficult to describe. A few seconds after the column began to rise, I felt the concussion and heard a loud boom, which was not anticipated by the scientists. The column of water looked foamy and was white with vertical sides. It rose to about six thousand feet above the lagoon with a cloud of steam at its top. From around the base of the column, I saw a white line which I thought was a small wave about five feet high. As the column began to fall, a circle of mist moving outward from the base of the column started rolling over the water like a strong wind, engulfing all the ships (Figs. 2A, B). The cloud dispersed after about fifteen minutes, but the mist was still too thick to see the ships, which were all thoroughly drenched in a heavy radioactive fog. The waters of the lagoon had also received a heavy dose of radiation, but the naval fleet returned to anchor within it and distilled drinking water from it. We even took seawater showers in it and were bathed in the spray when traveling in small boats from ship to ship and to Bikini Island for work duty. I thought nothing of it at the time, but shudder about the possible consequences when I think about it today.

What I had seen was my first *base surge*, years before I chose to become a volcanologist. This experience was incorporated into my career many years later in my study of pyroclastic flows and the base surges that sometimes form when water and magma mix.

Suggested Reading

Weisgall, J. M. *Operation Crossroads: The atomic tests at Bikini Atoll.* Annapolis, Md.: Naval Institute Press, 1994.

Figure 2. Nuclear explosion column and its collapse, Bikini, July 25, 1946. A. I was witness to the detonation of a nuclear bomb from its position ninety feet below the water of Bikini Lagoon. The explosion column, about half a mile wide, was an aerosol of radioactive gases and water droplets that rose to an altitude of about a mile.

B. The water-gas aerosol was heavier than air and therefore collapsed downward to the water surface. With no other place to go, the collapsed explosion column spread across the water surface at a speed of about sixty miles an hour, engulfing all of the target ships in a highly radioactive mist. (Photos: S. Glasstone, ed. 1950. *The Effects of Atomic Weapons*. Washington: U.S. Government Printing Office.)

21

◇◇◇◇◇◇◇◇◇◇◇◇◇◇◇◇◇◇◇◇◇◇◇◇◇◇◇◇◇◇◇◇◇

The Beginnings of a Career

The Earliest Calling

"Let's go to the hills tomorrow." Nearly every Friday, my friend Jimmy said that, and each time we usually found others to go with us, including my younger brother Bob. We even went when Jimmy couldn't. The hills were wondrous. For eleven- and twelve-year-old boys, these weekends were adventures of maturing freedom in the sage brush and grass-covered Puente Hills bordering the southern California town of Whittier (former president Richard Nixon's hometown), where I was born. The United States had not yet entered the European war that was raging, and it would be two more years before the Japanese engaged our country.

On these Saturday adventures, we took sandwiches and canteens of water and walked to the end of Citrus Street, which became an unnamed dirt path at the beginning of the hill slopes. The grassy hills were filled with hidden secret places, chaparral on the southern faces, and canyons with shady sycamores on the north. At some times of the year, a trickle of water ran in the creeks, though they were mostly dry throughout the year in southern California. We collected interesting stones and dug gypsum crystals out of the cracks of the earth. There were rare, huge tarantulas that would saunter slowly across the path, and an occasional snake.

As we sat for lunch one day, someone wondered, "How did the hills get here?" and "Why is the ground flat where the houses are built, down there below?" Those were perceptive questions far beyond our years of understanding, and none of us had ever consciously considered such questions. Raymond said, "The ground has always been where it is, and that is how it started." One said, "But don't things ever change?" And another said, "Probably not." I thought that maybe the hills were worn-down mountains. It was exciting to think that once there had been mountains looming thousands of feet above Whittier, maybe with snow on them.

In 1940, when I was in the seventh grade, my teacher, Mr. Kenneth Gobalet, introduced me to the excitement of many facets of science. I watched amoebae move across the field of view in a microscope. I stained one-celled animals on slides. I heated glass tubes on the flame of Bunsen burners and made glass bubbles by blowing into the tube. I learned that the mere touch of a match ignited phosphorous in great billows of smoke. One day I got a small piece of pure sodium and spit on it. Sodium is highly unstable in the presence of water, and I watched it dance across the chemistry table like it was a bead of water on a sizzling hot skillet. I touched the sizzling sodium with a cold knife blade, and it exploded. Dozens of small pieces of sodium struck my face like buckshot, barely missing my eyes; the next day I had chicken pox-like scabs all over my face. Mr. Gobalet calmly explained to me why it had exploded. He didn't scold me—he used the experience as a teaching device.

One day Mr. Gobalet told the class about Darwin's ideas on natural selection, about the evolution of physical and mental characteristics of animals, and hinted that all natural things evolve. I still remember my surge of excitement from this sudden enlightenment. Evolution! Everything evolves! It meant that people *could* evolve from apes and "lower animals" and explained why there are so many different kinds of birds. In my mind, I expanded the concept to mean that hills could evolve from mountains. It seemed like such a simple explanation that my twelve-year-old ego wondered why I hadn't discovered the idea myself. Mr. Gobalet had planted the seed; deep within my subconscious, my calling had become natural science, but it took several years for me to finally realize it.

The Beginning of a Career

My career began when I became a high-school dropout. I quit high school to join the U.S. Army shortly after World War II ended to get the educational benefits of the G.I. Bill of Rights. Because my mother and father had often urged me to, "Go to college," the message was deeply planted in my adolescent mind, but I had no idea what I wanted as a career. I intended for the G.I. Bill to finance my college career; high school was to be completed by passing a correspondence course while in the army. I entered college in 1948 as a music major because I had played piano in a high-school swing band and knew I could make a living as a musician. I had no idea that I could have a career in science. Through an elective course and quite by accident, I discovered the field of geology, a profession that I did not know existed. After a semester in geology, I changed my major and rediscovered the excitment of the natural sciences instilled in me by my seventh grade teacher, Mr. Gobalet. After four years, I received my B.A. degree with a major in general geology from Occidental College, Eagle Rock, California, in 1952.

Beverly and I were married in May 1947 before I was discharged from the army. She encouraged me to go to college, and together we decided to have a family. Our first child was born during exam finals in my sophomore year. We then decided to have a second child, and in my junior year, Beverly bore twins, a boy and a girl, during final examinations in 1951. We were a young, energetic and boisterous family when Beverly and I decided that I should go to graduate school for a Ph.D., because by that time I wanted an academic and research career in geology.

My study of *volcanic* processes evolved by chance, starting with my acceptance as a geological teaching assistant in graduate school at the University of Washington, Seattle, in 1952. There I met charismatic professor Peter Misch with whom I wanted to study, but whose overwhelming personality often dictated the research directions and results of his students. The day we discussed various topics that I might study, professor Misch mentioned, in his heavy German accent, "Once I traveled across the Cascades near Mount Rainier. Mount Rainier is a volcano built upon a basement of rocks made of older volcanoes. As I went west into the Puget Sound lowland I wondered how those old volcanic rocks related to the nonvolcanic sandstones of the same age. How does that transition occur?" That brief description was a pivotal point of my entire career, for the next day, after considering all of his projects, I picked that one, the farthest from his interests, to allow me freedom to tackle the subject on my own terms. The project was to discover the relationship between the older volcanic rocks of the Cascade Mountains and the light gray *sandstones* of the Puget Sound lowland, but the processes by which the volcanic rocks were formed were the most interesting to me.

Peter was disappointed by the project that I chose, but it started me on a lifelong study of volcanic rocks. Once, while I was discussing my thesis rocks with Peter, he said, "Richard, your rocks are the ugliest and most undistinguished rocks that I have seen in my thirty years in petrology." By that time, I had spent a month in the field and had found that most of the rocks in my dissertation area were of a type I had never heard about. There was no precedent for their study, and I was a very confused twenty-four-year old graduate student, but my curiosity had been piqued, and his remarks did not discourage me. To the contrary, they had the opposite effect, for I was dimly aware that I had a tiger by the tail.

The geologists that preceded me in the Cascades Mountains had written that the volcanic basement in my study area was composed of pyroclastic rocks, and this is what I expected to find. Although most of the rocks that I encountered were composed of fragmental products and were volcanic, I did not recognize any rocks that I could easily identify as pyroclastic, that is, fragmental volcanic debris blown directly from a vent. Instead, I found what I believed to be ancient mud-flow deposits containing fragments composed exclusively of volcanic rocks. I found layers of ancient river and lake deposits composed of volcanic fragments that contained abundant fossil leaves interlayered with some thin layers of coal, but they were definitely not pyroclastic rocks as stated by the experts.

I had stumbled upon an overlooked field of research—a large and diverse class of rocks whose origins were poorly described or not recognized. These rocks covered hundreds of square miles in the Cascade Mountains (as well as many other places in the world) and consisted of an accumulation of layers thousands of feet thick. One section of upturned strata in my area measured twenty thousand feet, nearly four miles thick! Moreover, there was no systematic precedent for the study of these diverse types of volcanic rocks, and, out of my utter confusion, I defined and classified them. This process identified some categories of volcanic rocks that opened several avenues for original studies.

Unrelated to my dissertation project, yet highly influential for my future career, was the presence of Mount Rainier, which greatly increased my fascination for volcanoes. "I love that mountain," I said to a friend one day. I had became attached to that massive old volcano, still active with a steam vent at its top, where climbers can spend the night in its warmth. Rainier was my silent and remote friend and served as a reference beacon. It dominated the landscape for hundreds of miles around and was plainly visible from Seattle, less than a hundred miles away. From nearly every ridge and peak in my area, Mount Rainier was in view, its crags towering above the landscape to show its snowy top and valley glaciers. I picked places for lunch with views toward the magnificent mountain, and overnight camp-sites within sight of the glowing light of the waxing and waning moon on its snowy cap, most spectacular during a full moon. Its beauty continues to hold my fascination.

Suggested Reading

Harris, S. L. Fire Mountains of the West; The Cascade and Mono Lake Volca-noes. Missoula, Mont.: Mountain Press Publishing, 1988.

◇◇◇◇◇◇◇◇◇◇◇◇◇◇◇◇◇◇◇◇◇◇◇◇◇◇◇◇◇◇◇◇◇◇◇

Central Oregon: Tuff, Fossils, and Lava

Ray Wilcox of the U.S. Geological Survey and I arrived in Monument, at that time a town of 150 people in central Oregon, on a sweltering hot, dusty July 4, 1957. Monument was a small collection of houses along the North Fork of the John Day River, where it emerged from a deep shadowed canyon that widened into a valley of open grazing lands and fields of alfalfa. The river had eroded through great stacks of lava flows known as the Columbia River Basalt. The deserted streets were lined with large shade trees, but we saw no one except an old man in a rocking chair on the porch of Boyer's Cash Store. We stopped.

Ray said, "Good afternoon," and the man, probably in his seventies, snorted, "Fourth of July and no one around. This is a dead damned town!"

That was my introduction to Monument. Fortune had smiled upon me once again and led me back that summer to central Oregon—a chance happening that permanently set the course of my research career onto pyroclastic rocks, ignimbrite, and volcanoes.

In 1955, before I finished my Ph.D. dissertation, I accepted a temporary two-year teaching job at the University of California at Santa Barbara. After finishing my dissertation in the spring of 1957, I was hired to continue teaching at Santa Barbara and applied for a summer job with the U.S. Geological Survey (USGS). By chance, I was picked by Dr. Ray E. Wilcox to assist him with a research project on *lava* flows of the Columbia River Basalt. The lava flow sequence overlies the John Day Formation in the very same region that I had visited in 1954 when I was working on my dissertation near Mount Rainier. The John Day Formation consists of tuff, and rocks of the formation are almost exclusively formed of pyroclastic particles erupted out of volcanic vents twenty to thirty million years ago. My luck went even further, for Ray Wilcox was one of the few world experts on pyroclastic rocks, and for two summers (1957 and 1959), I was his apprentice in central Oregon.

Ray Wilcox had driven the jeep from the USGS Denver headquarters, and I had flown from Santa Barbara to Redmond, Oregon, where he picked me up. We drove eastward that day through Prineville and Mitchell to meet with the John Day River at Picture Gorge, and then we turned north to Kimberly and eastward toward Monument. The scenery was breathtaking. In the broad valley were the tuff formations, magnificent red, green, and white eroded spires, castles, and pillars. In the distance at higher elevations were pine-covered, steep-sided mountains, formed from innumerable layers of stacked black lava flows one to two thousand feet thick (Fig. 3).

The land was vast and nearly devoid of people but full of the sounds of the river, the wind blowing through the trees, screeching hawks along the cliffs, and the songs of meadowlarks in the flat lowland meadows. The John Day River had carved a broad valley through the formations, revealing a sequence of rocks three thousand feet thick from the valley floor to the top of the mountains.

After our encounter with the old man in front of the Boyer's Cash Store, we rented two rooms for the night at the Stubblefield's house. Because there was no restaurant in town, we needed both room and board. The next day we found Wave Jackson whose occupation, in her words, was "Cookin' fer men."

Figure 3. The John Day Formation, central Oregon. The light gray castellated spires and columns are made of eroded twenty-million-year-old tuff. In the background, at a higher elevation, are several hundred feet of hardened black lava flows (Columbia River Basalt) stacked like pages of a book on top of the tuff. More than a hundred species of land-animal fossils, including primitive horses, have been taken from the John Day fossil beds. (Photo: Richard V. Fisher.)

Wave not only cooked, she washed our clothes, pulled the burrs out of our socks, and rented us an ancient shack with a creaking wood-plank floor in the back of her house. It had two iron-framed beds, a bare lightbulb hanging from the center of the room, and a small table consisting of an old door on boxes, with one old and unpainted wooden chair and a rocking chair on a small weathered porch. It was our home for two months. Wave also provided dinner entertainment. She regaled us with endless tales of the mores and social life of Monument, for she had been born there in the late 1800s and lived there most of her life. She had been a wild teenager, scandalizing the locals by riding astride her horse instead of sidesaddle. She continued to be a nonconformist, but Wave was our angel, providing us a place to stay and home-cooked meals, and taking care of the chores of daily living, thereby allowing us to spend two months fully immersed in the geology of the Monument quadrangle.

The John Day River now slices through the rock layers of the John Day country exposing them for study. For millions of years, the layers built up, as ash blew from the craters of Cascade Mountain volcanoes across eastern Oregon to settle on the grasslands of central Oregon. Gradually 2,500 to 3,000 feet of ash were deposited in former valleys, burying the scattered bones of primitive horses and other animals that once had grazed in that long-ago world, prey for long-dead carnivores. Today, vertebrate paleontologists discover bones in barren outcrops of tuff and can re-create from them the early animal population.

Few tuff layers are distinctive enough to trace across country, but one of the layers, an ignimbrite, can be followed across many valleys and mountains. The layer is two hundred feet thick at its maximum and covers 1,200 square miles or more of central Oregon. When I first saw it, I was awed by its persistence across many miles of landscape. Its great volume was astounding; it had to have originated from a catastrophic eruption!

The mornings were clear and crisp, and the afternoons hot and dusty. It was exhilirating work establishing the distribution of the John Day tuff, interpreting its origins, and determining from its composition that it resembled sources mostly from the Cascade Mountains one hundred miles to the west. Fieldwork during the summers of the late fifties and early sixties was so inspiring that I would sometimes hug a spire of tuff in my youthful exuberance.

One morning while I was walking across a canyon through the chaparral, a mourning dove flew up in front of me and then began flopping in the bushes as if it had a broken wing. Her nest, with four small eggs, was on the ground, half hidden in the brush. Just out of my reach, she continued to flop her wings, trying to distract me. She wanted me to believe that she was injured and would be a sure meal. I decided to follow her as long as I could. "Where are you going little mother?" I said aloud. I followed the dove for at least a mile until she suddenly flew away, having decided that her eggs were now safe. She flew in a direction opposite to the location of her nest and disappeared behind a ridge, no doubt to circle back to her nest. My admiration for the intelligence of living creatures was greatly enhanced by the encounter, and even today, when someone calls it instinct, I call it intelligence learned from God.

Another time after a long, hot day in the field, I sat silently in the shade of an outcrop with my legs outstretched, waiting for Ray Wilcox to pick me up along the gravel road. After a short time, there appeared a large black and white magpie whose occupation as a scavenger was to clean the pathways of roadkills. I remained still while the bird stood in the center of the dusty road for a few minutes with his head cocked first one way and then another. He hopped twice toward me and began to look at me again with his head cocked for a few minutes. He repeated his hop toward me and again surveyed me to see if I were a possible meal. The magpie was puzzled, for certainly I was not flattened and there were no blood or flies. He continued to get closer and closer until the hair stood on the back of my neck— this creature meant to eat me! I tolerated it as long as I could, and then, when he was a bird's length from my leg, I loudly said, "Not today, bird," and the bird instantly flew away.

Thanks to grants from the National Science Foundation and the University of California, I spent seven summers mapping the John Day Formation across about 1,200 square miles. The John Day rocks taught me many things about the region. First, it took about six million years to deposit the three thousand feet of tuff. This amounts to less than one one-hundredth of an inch, on the average, of volcanic dust per year, although geological processes do not operate on averages. Deposition is faster during some time periods, slower during others, and there could be hundreds of years during which no deposition occurred, depending on the size and number of volcanic eruptions and the direction and strength of the winds that blew the ash. In addition, except for the ignimbrite layer mentioned earlier, there are very few distinctive, widespread ash layers in the John Day Formation, suggesting that there were few catastrophic burials by ash. This fact alone meant that most of the time there was a stable living environment, and this is corroborated by a well-balanced animal population of herbivores and carnivores in the fossil record—gophers, mice, snakes, horses, and the now-extinct oreodont sometimes called a "pig-dog" a herbivore with canine teeth. Few if any heavy falls of ash had buried animals whole. Only isolated pieces are found—skulls, teeth, leg bones, ribs and the like—because the killers and the scavengers carried off and scattered various parts of the dead animals.

Third, I learned that the landscape during deposition of John Day tuff, sometime between forty and twenty million years ago, consisted of rounded grassy hills, lakes, and grassy plains. The climate at that time was temperate and much wetter than now because there were no high Cascade Mountains to shut off the flow of moist air from the Pacific Ocean at certain times of the year. There were active volcanoes off toward the Pacific Ocean but no continuous high barriers to squeeze moisture out of the clouds before they moved across central Oregon or Washington.

Most interesting to me, however, was the ignimbrite layer that was so pervasive across the area. It must have caused a colossal ecological disaster, for its deposits covered a minimum of 1,200 square miles of land that would have been devastated. (By comparison, the disastrous Mount St. Helens eruption in 1980 devastated about 230 square miles.) In many places, John Day ash particles became welded together to form a hard lava-like rock. That means it was extremely hot when it was deposited. Any animal or plant that existed in the region on the day of emplacement of the ignimbrite was blown away or buried and incinerated.

Recalling my experience at Bikini and the underwater explosion of the atom bomb that formed a base surge, I visualized the origin and emplacement of the ignimbrite layer in the basin to be similar to a base surge. In one of my articles published about the John Day Formation in 1966, I suggested that the pyroclastic material had exploded out of the conduit in a high eruption plume that collapsed back to earth and flowed along the ground as a pyroclastic flow.

Following the work on the John Day Formation and its ignimbrite, I became eligible for a sabbatical leave and arranged to spend a year in Hawaii (1965–1966). For a research project, I chose to work on a littoral cone named Puu Hou on the Big Island of Hawaii.

Suggested Reading

Fisher, R. V., G. Heiken, and J. B. Hulen. *Volcanoes: Crucibles of Change.* Princeton, N.J.: Princeton University Press, 1997.

◇◇◇◇◇◇◇◇◇◇◇◇◇◇◇◇◇◇◇◇◇◇◇◇◇◇◇◇◇◇◇◇◇

Puu Hou, Hawaii: Solitary Isolation

"It's a lonely place down there. No birds, no people. Nothin' there but black lava, goat grass, and that blip of a volcano, Puu Hou." I was in Naalehu, Hawaii; the Honolulu Ranch Company manager smiled and handed me the keys to the gate. The road down to the sea and on to Puu Hou started on the main highway that encircled the big island of Hawaii. I had started the research in 1965 while on sabbatical leave at the University of Hawaii, but returned to the mainland in 1966 before my work was completed. I had now returned to finish my project.

"Also, let me tell you, that road is the road of hell. Take a couple of extra tires. That road is just lava scraped by a bulldozer. Just pure rocky lava. That kind called *aa*. Rougher than hell and tears tires in nothin' flat. Not much of the smooth type *pahoehoe lava*."

"Don't expect to be there 'till late." He looked at the clock on the wall in his cluttered office. "It's four o'clock now, and it'll take you two and a half hours to get to the gate and down to the ocean. And it's only about ten miles. You can't go faster than five miles an hour."

It was October 16, 1966, the weather was foul, and I was alone. It was cloudy, and there were warm gusty winds and scattered rain showers as I slowly followed the road on the lava flow for two hours down to the edge of the sea near Puu Hou. The sky was going dark and the wind blowing strongly off the ocean when I arrived at my isolated destination on an island speck in the middle of the vast Pacific Ocean.

I aimed the lights of the jeep toward my chosen camp site and struggled to set up the flapping tent on the surface of an old pahoehoe lava flow, that smooth black lava rock that was once molten lava, about a quarter of a mile from Puu Hou. I had arrived one hundred years and six months after Puu Hou had been built, and I was there to see how it had been made.

All night the wind blew, and the rains poured down. Giant breakers crashed on the eroded edges of the lava flow, adding to the din of the wind, rain, and flapping tent. I wondered if the crashing waves could reach my camp site.

The night was pitch black as I lay in my sleeping bag feeling more isolated and alone than ever before in my life and worrying whether I would be able to complete the scheduled two weeks of fieldwork. I had traveled from Santa Barbara, and it was necessary for me to finish my study during the scheduled two weeks; I would not be able to return to Hawaii because the expense of this trip had consumed the last of my National Science Foundation research funds.

The Kahuku eruption, which led to the construction of Puu Hou, started on the night of April 6, 1868, on the southeast rift of Mauna Loa, Island of Hawaii, where Pele stamped her foot and a fissure opened in the ground at about 3,200 to 2,600 feet above sea level. Smoke and flames were not seen, however, until 6 P.M. on April 7 by a ship at sea. Lava that exploded when it entered the ocean at the seashore traveled eleven miles from the source and began to shower the land with fragments and construct the Puu Hou *littoral cone* at 9:30 that night. The eruption ended on the night of April 11 or morning of April 12; it had lasted only five days.

The few people who lived in the area at the time of the eruption first saw a glow that became brilliant orange streaks shooting skyward to outshine the brilliance of the stars. The lava burst forward with a hideous roaring sound along a fault line-turned-conduit. Magma poured from the great Mauna Loa. The volume of magma grew increasingly greater, and the streaks of molten lava merged to form a great roaring, glowing, orange-hot curtain—a "curtain of fire," as it is called—a familiar sight over the centuries to the native people of the island.

People lived nearly a mile down the side of the mountain where it slopes to Ka Lae (South Point), which is now the southernmost point in the United States. The lava began to pour out of the great fracture in the ground and, slowly at first, moved down the grassy slope toward Ka Lae. The uneven landscape, however, caused the streams of lava to diverge and then join together again to leave small green areas of the landscape (kipukas) surrounded by a black sea of lava. The few residents had time enough to evacuate, because lava flows generally move slowly enough to allow people to run away from them. The lava then found a steeper way to the sea at the base of the Kahuku cliff (a fault escarpment). The steeper slope caused the lava to move faster, and soon all of the lava from the source was flowing at the base of the cliff toward the sea.

After five days, the lava stopped pouring from the vents, the flows stopped, and the explosive activity at Puu Hou died away to hissing steam vents as the lava slowly cooled. What was left were four mounds of fragmental material the size of small volcanoes. The highest, about 250 feet, was named Puu Hou (New Hill). Because they had cone shapes and originated in the littoral zone along the edge of the sea, they became known as *littoral cones*. They looked like volcanoes but had no underground vents.

After my second long and windy day of work on Puu Hou, I heard a shocking radio announcement headlining the six o'clock news, just as I was opening my can of stew for supper. Because of a large earthquake in Chile, there was a *tsunami* alert for all of the Hawaiian Islands.

I was stunned, for I was camped within fifty feet of the edge of the sea and only fifteen feet above it. I had read about some of the destructive tsunamis that had occurred in the past on the island of Oahu and the one at Hilo in 1946 that had destroyed buildings along the waterfront and inland for about a hundred yards or more. I tossed out my stew, took down the tent, and threw all of my belongings into a jumbled pile in the jeep, and drove as fast as possible up the scraped lava road for about a mile until I was about two hundred feet above sea level. I tried to sleep, but it was cramped in the backseat of the jeep. I couldn't set up camp because the aa lava was too rough. Moreover, it was raining. I spent a sleepless night listening to the radio and the rain on the roof of the jeep. Leaving my seaside post had been needless, for it was announced that "At 11:52 P.M., a four-inch tidal wave hit Hilo." I returned the next morning to my campsite.

After setting up camp, I tried to establish a routine. I awakened as the night sky turned to morning, drank a cup of cold coffee for the caffeine, opened a can of peaches, and sat outside the tent to eat them. Within minutes after opening the can, I was visited by a large black bumblebee. Thereafter, I ate the peaches first and then went outside to drink my coffee and breathe the morning air. The other living beings that I saw during my stay were a spider, dolphins, and a goat far in the distance across the lava flows—a white patch on a black background.

My tent was on barren pahoehoe lava that had flowed off Mauna Loa long ago. It was located close to the mounds on which I did my work, and the edges of Puu Hou littoral cone were directly on the old flow. After the day's fieldwork and a dinner of canned beans or hash, fruit, and a cup of instant coffee in cold water, I would sit on my folding chair in front of the tent. (I did not bring a stove for cooking because it required gasoline.) I worked on my notes and would watch and listen to the waves crashing on the barren rocky shore (Fig. 4). Occasionally I would search the horizon for boats, but none were seen.

Several late afternoons, at about five or six o'clock I was entertained by dolphins. Some would leap straight out of the water, twirling and twisting, and then flop on their sides onto the water

Figure 4. Campsite on Hawaii's dry south coast, October, 1966. Evening review of notes at my tent, which sits on smooth pre-historic pahoehoe lava. The ocean breakers are crashing on hard lava at the shoreline fifty feet south of the tent (to the left). In the background, Puu Hou littoral cone stands about 250 feet above sea level. It grew to that height in just five days: lava flowing into the cold waters of the Pacific Ocean caused hundreds of explosions that sprayed the land with sand- to football-sized pieces of chilled, solid lava. Waves have cut a cliff into the south side of this littoral cone. (Photo: Richard V. Fisher.)

with great splashes. They played games of catch-me-if-you-can or follow-the-leader. Watching such performances was like attending the ballet.

I would also watch the beautiful red sunsets. In the warm and windy evenings, when clouds did not obscure the sky, I would watch the stars wink on one at a time. When the sky went black, the crystal-clear air revealed even the tiniest stars, extending as a hemisphere down to the horizon of earth and sea. I lived in a hemisphere, like a glass Christmas snow globe. At times I would sit outside my tent to watch the stars in the October sky. On clear nights, the waxing moon revealed every detail of its face; by October 23, it was about three-quarters full and tinted the dark coast in dim highlights and deep shadows.

After the second night at Puu Hou, my two flashlights had burned out, so I slept according to the darkness of the night. In the darkness, I would crawl into my tent and into my sleeping bag, to be lulled to sleep by the incessant roar of breaking waves and the smell of the salty air and be awakened by the sunrise.

My days were solitary but indescribably peaceful. I was alone but not lonely. My mind was fully occupied as I busily mapped the distribution of the black *basalt* lava as well as the mounds of exploded lava scraps (basalt ash, basalt *lapilli*, and basalt *bombs*) that formed when the fronts of the lava flows entered the water and exploded, spraying back steam, water droplets, chunks of congealed lava, and hot lava droplets. Some liquid droplets were as large as human heads, and when they fell to the ground the outer chilled margins cracked upon impact to form surfaces like a bread crust.

The wind blew constantly, and the relentless breakers along the coast crashed to form sea spray that filled the air with a nostalgic fragrance of childhood summer picnics with my family at Long Beach, California. At times I sat on the littoral cone and faced the ocean to feel the spray carried by the breezes, and dreamed about the life of Hawaiians before the coming of the Western world.

Four of the days of my two weeks were spent exploring the congealed lava rivers that once flowed into the sea. After breakfast those mornings, I filled my canteen with water and threw a can of beans in my backpack, along with my camera and sample bags, and followed the 1868 lava flows back up the slope.

The hundred-year-old landscape was similar, I imagined, to the early years of the earth. The total silence—except for the pinging and knocking of my geology hammer against the rocks—was like a warm blanket of solitude. There were absolutely no signs of human habitation nor of civilization. I was in a world seemingly without living creatures. As I followed the confines of the lava river, I saw no birds or insects. Many times before, I had worked and walked alone through valleys in the forests of Oregon and Washington, in dry, sandy river beds of the chaparral-covered mountainous lands north of Santa Barbara, and through hot, winding, dry sand beds of desert lands, but never had I felt so safe as I did in the empty land along the southwestern coast of Hawaii. There were no poisonous snakes, no dangerous animals, and no people. The only danger was from myself—a rock chip in the eye, a sprained ankle or broken leg, a fall into a cavernous lava tube through its ceiling skylight. I had spent many hours alone in the field over the previous twelve years and practiced careful safety procedures—"never take a chance no matter how small." Walking within the corridor of basalt, sitting to contemplate the meanings of physical shapes and relationships of lava patterns, and writing observations in my notebook in such quiet isolation in the warm sunshine was total peace of mind.

In the channels, I saw where red, boisterous flows carried chunks of lava as large as Volkswagens, rounded as they had rolled down the lava channel. At times I stood in awe of the tremendous forces required to produce what I was seeing. Slabs of black basalt had cracked in their centers, resulting in protrusions of lava that had forced their way upward from the liquid lava beneath the lava slabs. In other places, there were rough-and spiny-aa lava lobes that had opened at their snouts, like broken eggs; fluid basalt had flowed out and congealed as smooth pahoehoe. When I closed my eyes, I imagined the roar and cacophony of lava slabs carried by liquid lava, smashing against one another in great confusion as the black river, spouting from the side of Mauna Loa, flowed toward the sea.

I had read about lava flows that formed high-standing levees paralleling the rivers of lava, but I had not seen them up close. As the lava flows, it continues to cool at its surface until there

41

are solid crusts floating on the mushy red hot lava. Because the fluid lava moves faster than the solidified crusts, the crusts break into slabs that catch on projections from the sides of the channel and slow down. The slabs are then swept to the sides by the faster-flowing liquid rock, and so they begin to pile up to form levees, higher than the flow, along both sides of the channel, even if the flow starts on a flat surface. Occasionally, the volume and speed of lavas increase, causing the flow to rise between its levees and then subside, leaving a plaster of black basalt coating the fragments. Levees can build up to twenty or thirty feet above the bottom of a channel, but the bottom also may build up as the lava congeals at its base. Lava makes its own confined channel that grows in height as it moves down a slope.

My eyes were so attuned to looking at the details of the lava and lava forms and interpreting their origins that I didn't see the myriad life forms surrounding me until a glint of light captured my attention. I watched what appeared to be a snippet of hair brightly reflecting the light from the sun. The apparition floated slowly toward me on a seaward-moving breeze until it was but a foot from my eyes. I watched it float downward and catch upon a strand of grass that grew in a crack in the lava. As I knelt down to look more closely, I discovered a spider no larger than the period at the end of this sentence, and once again marveled at the tenacity of life. If there was one spider, there must be many more: Tiny bodies clinging to life on strands of grass in a giant sea of crystallized lava. Looking at that tiny speck of a spider, I felt a kinship, for I, too, was a very tiny being on this planet called earth, itself a small speck in the vast universe. "We're in this together," I said to the spider.

One morning I found out about the intelligence of goats. Near the end of one of the lava flows, there was a mound that looked like a small volcano about twice as high as my rented jeep. I wanted to see how it had formed. It stood about two hundred yards from the edge of the flow within a sea of extremely rough aa lava with many ridges and depressions, but I decided that I could get to it. Two hours later, at ten o'clock, I arrived at the cone after having carefully picked my way over the razor-sharp surface of loose aa lava fragments, constantly sliding and stumbling my way down into the depressions, climbing and backsliding over the small ridges. The cone had been formed by hot, gas-rich lava that had blown through the crusty surface of the lava stream and fallen to earth fast enough and fluid enough to form small lava flows. The spattering lava was piled high enough to form a tiny volcano form known as a *spatter cone*.

Satisfied with my discovery and interpretation of how the cone was formed, I carefully began to make my way back, dreading the return. As I started, I saw a tiny whitened goat dropping clearly visible against the backdrop of black lava and then immediately saw another about twenty feet farther on. I guessed that the two pellets were the sign of a trail left by goats, and I started in that direction. The surface was still rough, but the depressions and ridges were not as severe, and when I had traveled twenty feet to the second pellet, I saw a third, a fourth, and so on. I followed the pellet trail back to the edge of the lava flow. It had taken two hours to make the trip to the cone but only twenty minutes to return!

When I finished my study of Puu Hou, I felt that I understood how littoral cones were made. I found that they are mounds of debris constructed by explosions from the interaction of water and hot lava at the point where lava enters the sea. At Puu Hou, the lava entered the sea amid thunderous explosions and the generation of great volumes of steam. Explosions occurred where fast-moving lava was trapped beneath colliding slabs of solid lava piled on one another and seawater was encountered in a confined environment beneath the jumble of slabs. Water suddenly heated and trapped in a confined space by lava at a temperature of about a thousand degrees Celsius expanded quickly and violently. Explosions ripped solid lava rafts and molten lava into pieces that were flung into the sky simulating a volcanic eruption. The fragments rained down to make the mounds at the end of the lava flows. One mound was Puu Hou, with its seaward front forming a cliff of basaltic tuff.

43

The littoral cones lack feeding vents connected to subsurface magma supplies (they are "rootless") and form where lava moves over small ponds of water, swamps, springs, or streams, or where lava enters the sea. Explosion centers are near or at the shoreline; therefore, about half of the radially exploded material falls into the sea, leaving a half-cone on land. A typical littoral cone has a wide crater (if the part missing at sea is reconstructed) and low rims, steep inner slopes exposing the strata from which they are made, and gentle outer slopes merging with that of the underlying terrain.

The primary circumstances necessary for the construction of littoral cones appear to be the rapid delivery of large volumes of lava to the water and confining conditions in which water and lava become repeatedly mixed and the pressure of expanding steam blows apart the lava. The abundance of layers within littoral cones and the height to which they can be built perhaps to three hundred feet) suggest that these conditions occur repeatedly.

Satisfied with the data that I had gathered, and carrying close to a hundred pounds of samples in my backpack, I started for home. The agricultural inspector at the airport wanted to check my rocks to see if there were any soil samples, and I told him they were all lava and cinders. He looked at me solemnly and said,

"You know that you are taking a risk."

"Too much weight?" I asked.

He answered, "I know several people who took small rocks out of Hawaii, and they had much bad luck. You have at least a hundred pounds. It is a very serious matter."

"Is it legal?" I asked.

"Well, yes," he answered, "But Pele is hard on those who carry lava away from the island." In my heart, however, I knew that Pele would approve of work that honored her creation of Hawaii.

I returned to my laboratory in Santa Barbara to study the specimens, analyze the ash, and write my research paper, which was published in 1968, exactly one hundred years after the eruption. My curiosity about the effects of the interaction of water and hot lava was whetted, and it led me to apply for another research grant, this time on the effect of water on the formation of maar volcanoes.

Suggested Reading

Stearns, H. T. Geology of the State of Hawaii. Palo Alto, Calif.: Pacific Books, 1966.

Chapter Six

◇◇◇◇◇◇◇◇◇◇◇◇◇◇◇◇◇◇◇◇◇◇◇◇◇◇◇◇◇◇◇

Volcanoes and Water

When I was twelve years old, I read every book I could find written by Richard Halliburton, and in my mind I traveled with him to exotic places all over the world: Tibet and Mount Everest, Africa, China, and more. Halliburton whetted my desire for travel, but I had no expectations that it would be fulfilled. My study of Puu Hou, however, was a gateway to the world. It was the catalyst for my developing curiosity about the origin of Hawaii's *maar* volcanoes near the edge of the sea, but maars occur in many parts of the world, and eventually my research carried me to Argentina, China, France, Germany, and Italy, Japan, New Zealand, and the Philippines.

My earliest encounter with maar volcanoes was in September 1965, while on sabbatical leave at the University of Hawaii to study littoral cones on the big island. I explored the rim and slopes of Diamond Head, Oahu and was transfixed by the shallow flat-floored bowl-shaped crater that was over half a mile in diameter (Fig. 5). I was also intrigued by the yellow-brown color of the volcano. The color results from the abundance of *palagonite* that forms the rock layers. Palagonite is a yellow-brown mineral altered to clays and oxides from black basalt glass, which is formed when magma or lava is quickly quenched by water. The interaction of basalt magma or lava and water is often explosive, forming abundant tiny glass fragments deposited as ash. It fol-

Figure 5. Diamond Head, Hawaii. This bowl-like crater is a typical maar volcano, constructed from numerous shallow explosions by molten magma in the
volcano's conduit mixing with the water as it rose into a coral reef at the edge
of Oahu, Hawaii. The more familiar view of the profile of Diamond Head from
Waikiki beach shows a rim that slopes upward to the south and stands notice-
ably higher at the water's edge. This slope is thought to be the result of strong
winds that blew south during the eruption. (Photo: NOAA.)

lows that the presence of abundant palagonite in the tuff of Diamond Head means that it was originally built of abundant basalt
glass fragments. Another indication that water was present during eruptions that formed Diamond Head is the abundance of
coral reef fragments imbedded in the tuff. It shows that the volcanic eruptions exploded through a coral reef in the water at the
edge of the island.

I stopped at various places along Diamond Head's crater rim
to enjoy the trade winds blowing across the crest and wonder
about the volcano's origin. Why was the crater was so wide
compared to its height?

I also visited Koko Head east of Diamond Head and happened to encounter some thin ripple- and dune-shaped layers. I was puzzled but did not question why they were on the side of a volcano. Without critical thought, I assumed that rain may have formed streams that carried and rippled the ash down the sides of the volcano and then promptly forgot the issue. I did not understand what the rocks were telling me.

Named from the Latin word *mare,* meaning sea, maars were originally recognized as small, nearly circular crater lakes in the volcanic district of the Eifel region of western Germany. When I returned to Santa Barbara from Hawaii in the fall of 1966, I began writing a grant proposal to study maar volcanoes and invited Aaron Waters, a well-known leader in the study of volcanic rocks, to join me in a joint research effort. He was eager to participate because he wondered about the origin of palagonite in maar volcanoes. At that time, there was a controversy about how palagonite was formed from basalt glass—whether it formed at the time of eruption from hot ash or whether it was formed over many years by atmospheric alteration processes as water seeped through the ash. In 1966, NASA (National Aeronautics and Space Administration) was funding research about the moon. Because of the imminent moon landings, we applied to them for funding. We proposed a study of the shapes of maar volcanoes as moon analogues. We argued that if some of the craters on the moon are maar volcanoes, then there must be a layer of ice beneath the moon's surface as a source of water. We were awarded a grant from NASA in 1967 to investigate the origin of maars. It turned out that our premise was completely wrong, for rocks brought back from the moon in 1969 were physically and chemically dry. Bone dry. Instead, it was revealed that the craters on the moon are formed by impacts with asteroids and other space debris.

Another place the research took me was New Zealand. The country's largest city, Auckland, is built on the products of basaltic volcanism with numerous *cinder cones* and maar volcanoes. I arrived there in June 1969 and was there when Neil Armstrong first touched the surface of the moon. It was the afternoon of July 21 in New Zealand, and I had traveled south from Auckland that day to investigate one of the many maars in the region. I happened to have a handheld portable radio with me, and, as the event took place I sat on the rim of a maar overlooking the grassy sheep pasture of the crater, with my radio pressed tightly to my ear. The thrill of involvement with such a monumental historical event was incredible: Here I was on a NASA research project, sitting on the rim of a maar crater, a possible moon analogue, and listening as a human being touched the moon. When Armstrong touched the moon, at exactly 2:56 P.M., I simultaneously clapped and danced a jig on the earth. It was July 20 at 7:56 P.M. in Santa Barbara, California.

Maar Volcanoes and Base Surge Deposits

I started dedicated fieldwork on maar volcanoes in the summer of 1967. In June, I visited Grant Heiken, my first doctoral student in Santa Barbara's newly instituted Ph.D. degree program in geology. His dissertation research focused on maars in south-central Oregon, and I wanted to see what he had found. I also brought him a newly published paper by Jim Moore of the USGS about the eruption of Taal volcano in the Philippines and the *base surges* that it had produced.

The base surges at Taal had deposited ash ripples and dunes from the volcanic winds, and the layering of the deposits showed undulating structures like those I had seen at the Koko crater in 1965. The realization by James Moore that the horizontally moving eruption clouds seen at Taal volcano in 1965 were analogous to those seen at Bikini in 1946 (chap. 2) had finally shattered the dam of ignorance, even though Adrian F. Richards was first to point out the similarity in his 1957 Ph.D. dissertation. By linking the formation of volcanic ripples and dunes to base surges at Taal, it was at last possible to understand the origins of ripple- and dunelike bedding on maar volcanoes such as

49

Koko Head. After Grant read Moore's paper, he showed me similar dunes in his dissertation area.

The story behind the discovery illuminates the serendipity of research directions, which generally follow the whims of curiosity. Jim Moore was in Japan when Taal volcano erupted in mid-September 1965. He hurriedly got permission from the USGS and the Philippine government to observe the eruption. Moore, Kazuaki Nakamura (University of Tokyo, Japan), and Arturo Alcaraz (Commission on Volcanology, Philippines) published a short paper in 1966 on the eruption in the journal *Science*, noting that "at the base of the main cloud column a flat turbulent cloud spread out, radially transporting ejected material with hurricane velocity" (p. 955). Moore was fascinated by the horizontally moving cloud and continued to investigate the phenomenon. He thoroughly researched the literature and discussed the Taal eruption with his colleagues. One of his friends happened to be a reviewer of my paper on the origin of pyroclastic flows submitted for publication on October 15, 1965. In that paper, I had theorized that pyroclastic flows could be formed by a collapsing column like the Bikini base surge. Moore sent me a letter dated December 15, 1965, asking "Could you please send me a reference? [Anonymous] mentioned seeing a reference dealing with *effects of Atomic explosions* in a recent paper of yours he reviewed. I am interested in referring to it in connection with work I am doing on the Taal eruption report and the generation of the horizontally moving eruption clouds." Science cannot progress very rapidly if communications are not open, and I gladly sent him the reference, and in doing so, I did myself a favor. The knowledge that I gained from Moore's work opened the door wide for my own research on base surge bed forms and their importance in interpreting the origins of pyroclastic deposits.

Although his work was initially overlooked, the first geologist to note that a volcanic eruption resembled the base surge at Bikini was Adrian F. Richards in his Ph.D. dissertation research of Bárcena volcano on the island of San Benedicto, Mexico, submitted in 1957 to the University of California, Los Angeles, and published in Bulletin Volcanologique in 1959. He stated that "The most striking feature of the birth of the volcano was the behavior of the eruption column. Its horizontal spread increased much more rapidly than its vertical rise and resembled the base surge of an atomic explosion" (Bulletin Volcanologique 1959, p. 89).

Zuni Salt Lake Maar

In late August of that summer, my youngest son Peter, who was ten years old, and Steve Donnel, a friend of his, went camping with me in California, New Mexico, Arizona, Utah, and Nevada on a trip to examine several maars of the western United States. I especially remember the maar at Zuni Salt Lake, New Mexico. It was a hot August 17 when we stopped at the Zuni pueblo to ask about road conditions to Salt Lake. "It's a long way through beautiful and sacred country. Treat it like your mother," said the man at the trade center.

We ate some fry bread there, a real cholesterol treat. The native people take dough, deep fry it, and serve it with honey or other toppings, such as beans. We then started on our way. Ten miles east of Zuni, we turned south on State Highway 36 and drove thirty-two miles through beautiful and unpopulated juniper-covered land to a few buildings called Fence Lake. We then turned onto a poorly maintained dirt road that would take us to Salt Lake. It had rained heavily a few days before, there were places where the road was rocky creek beds, and we almost got stuck in a sandbar. The map said twenty miles to Salt Lake. I had expected to see a familiar shape rise above the land, but there was still no mountain where the volcano ought to be.

About fifteen miles from Salt Lake, the land began to rise on a very gentle incline and the road became sandier. At the twenty-mile mark, we came to a gate with a sign that read, "N.M. Land Comm'r Lessor Southwest Salt Co." I drove through the gate and up the dusty road for a hundred yards or so and suddenly we were inside the crater of the volcano, near the salt works. For the past five miles, I had been on the outer flanks of the volcano, but the rise of the outer slopes was barely perceptible, and certainly not conspicuous enough from a distance to make one think that there was a volcano present. The crater, one and a half miles in diameter, has an essentially flat floor occupied by a small lake on the north side and two cinder cones in the center. The eruption had blasted a crater through the ancient sandstone that crops out over the region. The floor of the crater is about 150 feet below the general level of the surrounding landscape. Volcanic ash derived from the eruption is thickest on the rim and gradually thins out away from the volcano. The highest part of the present-day rim is about four hundred feet above the crater floor.

I approached the office building where there were several high mounds of salt ready to be transported. The dried parts of the lake were being dredged for evaporite salt deposits. Mr. Hagen, who ran the place with his wife, was inside.

"Going to be a hot one today," he said. "What can I do for you?"

"My name is Richard Fisher, and I'm from the University of California at Santa Barbara doing research on several volcanoes in New Mexico. Could I have permission to walk across your property?"

"Sure can," he said, "but this isn't much of a volcano. There's no lava, and there's nothin' here but a salt lake and those two small cones in the center of the lake basin don't amount to much."

"Today I want to visit the two cones in the center and look around at the cliffs." I explained. "And I've got my son and his friend with me. That OK?"

"Be my guest, just don't leave any papers and cans around and no shooting in the lake area. And you're welcome to camp anywhere around. But," he said with a twinkle in his eye, "unless you're part Indian, you could be visited at night by the ghost of the Apache warrior. There's been several sightings. He

Figure 6. The central crater area of Zuni Salt Lake maar, New Mexico. The crater contains a lake and two small cinder cone volcanoes. The low white cliff (fifty feet high) in the background is made of nonvolcanic sandstone layers through which the explosion occurred; it is the north rim of the volcano. The photograph was taken from the south rim. (Photo: Richard V. Fisher.)

doesn't like white men because they've ruined his country and killed his people."

"Is this Apache country?" I asked.

"Not exactly, but that's no problem for a ghost."

"Well, I'll tell the boys about it, and many thanks for your permissions," I said.

I parked the van outside the gate because they closed at 5 P.M., and we started for the two cinder cones, skirting around on the west side near the rim (Fig. 6).

I spent the rest of the day taking pictures of the cones and looking at the rocks. I also looked at the basement rocks of solid sandstone and could scarcely imagine the gigantic forces involved in blasting a hole one to two miles in diameter and fifty to one-hundred feet deep through solid rock. I was most interested, however, in looking at the deposits of the rim beds of the volcano the next day.

We got back to the van late in the afternoon and searched for a place to camp. I decided on the eastern rim of the volcano because the ash was soft like beach sand, and because it overlooked the crater and the magnificent sunset that evening. I set up our small tent; then I set up the stove and made hamburgers. The boys found some wood, and we made a small fire to sit around until bedtime. I could hardly wait to tell them about the ghost Apache warrior.

The following two days, with my raised awareness of maar volcanoes, dunes, and base surges, I was able to see many features in the rock layers that I had never observed before. The best exposures were on the south and north rims and showed example after beautiful example of *cross beds* (layers truncated by other layers across eroded edges) in dunelike bed forms. Once, when the boys were not fooling around elsewhere, I pointed out a small outcrop that had ten thin layers, no more than half an inch thick each, slanting forward on another layer that was horizontal. At the top of the ten layers, another layer clearly truncated them at an angle.

"How do you think that the layers got this way?" I asked.

"I dunno," Peter said. He clearly wanted to spend his great exuberant energy some other place.

"What about you, Steve? What do you think?"

Steve obviously wanted to say something more intelligent than Peter. He said, "I think that the volcano did it."

"That's right!" I said enthusiastically. "But how did it do it?"

"Look," I said pointing to the truncated edges of the ten thin layers, "What happened to the other end of these layers? Why does the upper layer cut right across them?"

They looked, definitely wanting to be elsewhere looking for lizards.

54

"I'll give you a hint. The wind blew across the land, leaving layer after layer of volcanic ash. What happened to the ends of the layers?"

"They blew away," Peter said gleefully. "Can we go now?"

I said, "Where did the wind come from? Look, the ends are cut off on the side toward the crater."

"It was a windstorm."

"From the crater? Do winds come from volcanoes? Or do they come from the sky? This one looks like it came from the volcano."

The boys were not impressed. I told them to go play, and they thankfully ran down the gully.

"Slow down," I yelled. "Watch out for rattlesnakes. Remember the one we saw this morning!"

I would not have been able to follow the line of questioning with the boys that morning had it not been for the eruption of Taal Volcano in 1965. Prior to that, the dune forms and cross beds seen in maar volcanoes had been attributed to deposition of ash redistributed by water or wind after the layers were deposited. The prevalent idea of the time was that, as the ash fell, the atmospheric winds blew.

Japan and the Philippines

One of my research studies of maars took me to the Philippines in June 1970 to examine the dune structures and cross beds in Taal Volcano, for that had been the only modern eruption in which base surges had been observed, with dune deposits developed from them. On the way, I visited the maars on Izu-Oshima, and island in Japan, to compare them with the ones I had worked on in the western United States. I found that their structures were alike, suggesting similar constructional processes as basaltic eruptions encountered copious amounts of water.

I had contacted Dr. Naoki Isshiki of the Japanese Geological Survey, whom I had met previously in Santa Barbara, and he promised to show me the maars on Izu-Oshima, a volcanic island with many hot springs in Sagami Bay about sixty miles southwest of Tokyo. I met Isshiki at the harbor in Tokyo, and we waited in a large barnlike building full of Japanese people going to Izu-Oshima for a holiday. The boat that we took was built before World War I, and when it left the harbor at 10:00 P.M., a band on the waterfront played Auld Lang Syne. We arrived at the island at 5:00 A.M. and took a tiny shuttle bus to the Japanese-style hotel, Kawaki-En. We rested on floor mats until 6:30, and then I went to my first Japanese breakfast. It was buffet style and included many kinds of raw fish, boiled and raw octopus, raw eggs, unrecognizable green vegetables, and seaweed of several kinds. I served myself the Japanese foods but then decided to have a dollop of mashed potatoes just to remind myself of home. When I tasted one of the green vegetables, I found it to be so bitter that I could barely eat it. Not knowing Japanese customs, I asked Isshiki, "Must I eat everything on my plate?" He curtly said, "Yes!" I felt sick but decided to eat the bitter greens, "a delicacy of Izu-Oshima," with my mashed potatoes, so I put greens on my fork with a bite of potatoes. My sinuses exploded, and my eyes watered—the potatoes were horseradish. I decided to mask the bitter taste of the greens with horseradish and was able to finish them. The raw fish of various kinds proved to be the better part of the breakfast, but as we walked out, I noticed that there were many things left on customer's plates. I mentioned it to Isshiki, and he said curtly, "Yes!"

We spent a rainy day visiting many outcrops of maars and ash-fall tuffs around the island, observing dunelike layers within outcrops of maar volcanoes. They were the same forms I had seen in Hawaii, central Oregon, and New Mexico. That night, we stayed at a small Japanese hotel—a large room with eight tables and a mat floor—where Isshiki had spent many nights while working on his doctorate degree. It was run by a tiny old woman who did laundry and, for a small fee, would allow people to rest for the night. While we were there, we took off our wet and muddy pants for her to wash and iron while we sat in our underwear, drank tea, and talked about maar volcanoes. The next day I continued my journey to the Philippine Islands.

As the plane arrived in Manila on June 21, 1970, the pilot announced, "The humidity is 90 percent; the temperature is ninety degrees," and it was nearly unbearable. In the air-conditioned hotel, I contacted the Philippine Volcanological Commission (VOLCOM) and made arrangements to go to Taal Island within Lake Taal. They couldn't take me to the observatory until June 25, so I hired a driver for four days to take me to Lake Taal, visit Richard Taylor, my brother-in-law stationed at Clark Air Base, buy needed topographic maps from a government agency, and search archives at the newspaper office for photographs of the eruption in 1965. The slowness of the bureaucracy at every transaction stretched out the time so that I just finished my business when it was time to go to Taal Island.

The first night at Taal, I had a dinner of baked chicken and rice, a Philippine staple, with the technicians who manned the volcano observatory. It was a very warm welcome to the Taal contingent of the Volcanological Commission. The dinner was excellent, but the next day, my first day of research on the island, I contracted a stomach ailment complete with diarrhea. I felt miserable and could hold nothing in my stomach. There I was, thousands of miles from home, having invested my grant money in the trip and with research to conduct. So I did. In my misery, I recalled that Charles Darwin was constantly seasick on his ocean journeys to South America.

The technical staff at the observatory were six men whose leader was Contrado Andal, the Volcanology Observer with the Philippine Volcanological Commission. He operated the volcanological station at Barrio Alas-as on the west-central coast of Volcano Island and witnessed the eruption of Taal in 1965. At that time, the observatory station was located about half a mile west of the 1965 vent. Before going to bed at 11 P.M. he had not seen any record of tremors on the seismograph, but when he awakened on the morning of the eruption, there was a record of continuous vibrations. Looking toward the volcano, he saw incandescent materials being ejected toward the southwest at about a forty-five-degree angle. Because of the direction of eruption, he advised his neighbors to evacuate the island and head north. His boat normally held six people, but he took twenty, and saved them all. When Contrado returned an hour or so later, the station was still intact. A day later, it was buried under ten feet of ash and boulders.

The eruption began as an ordinary *Strombolian* type (an incandescent lava fountain that would have constructed a cinder cone), except that a crack formed on the side of the volcano facing the lake, thereby allowing water to pour into the vent. The water mixed with the rising magma, which radically changed the behavior of the eruption from Strombolian to hydrovolcanic. The resulting explosions produced base surges consisting largely of steam and resembling the one that I had seen nearly twenty years before at Bikini (Fig. 7), but the Taal base surges were full of ash and larger fragments of volcanic rock. Like hurricanes, they swept over the landscape, tearing trees from the ground and sandblasting the trunks of those left standing. They flowed across the water and engulfed fleeing boats. About 190 people were killed by the Taal base surges.

Fragments large and small rained down from the turbulent base surges, leaving a carpet of deposits. For the first time, it was seen that a volcanically produced, horizontally moving, turbulent cloud composed of abundant particles can form dunes somewhat similar to desert sand dunes but more streamlined. The big difference between base surge dunes and desert dunes made by winds is the presence of fragments too large to have been carried by ordinary winds.

Roger Datuin had been assigned by the director of the Volcano Commission to accompany me into the field. Taal volcano had erupted several times in 1965 and 1966, each time with base surges that formed dune fields and plastered wet ash onto still-

Figure 7. Taal volcano, Philippines, in eruption, 1965. Lake water poured into the crater and mixed with rising magma to create a water-rich explosion column, a mixture of exploded bits of lava and steam. The column collapsed to the ground, and the turbulent mixture moved across the land with the force of a hurricane. The explosion resembles the base surge created by the underwater detonation of an atomic bomb at Bikini Atoll (see fig. 2B). (Photo: Courtesy of Manila Times, Manila, Philippines.)

standing trees. But it erupted yet again in 1968, this time Strombolian style to form a cinder cone within the 1965 explosion crater. On my first day in the field, Roger and I explored the source area of the eruptions and climbed the cinder cone streaked with yellow sulfur spots. Although it had been two years since the eruption, the cone was still fuming at several places. The ground was warm, and the sulfur-dioxide fumes (like those that come

Figure 8. A field of dunes left by base surges from the eruption of Taal volcano, Philippines. For the first time, the cause of ripples and dunes composed of volcanic ash and rubble were explained. Previously, they were thought to have been shaped by water or ordinary wind. (Photo: Richard V. Fisher.)

from striking a match) were noxious in some areas. The experience was surreal—to be in a tropical land standing on a cinder cone surrounded by a nearly barren desert.

With an amoebic disorder, I felt terrible and was poor company, so I worked alone. Some of the features of the Taal base surge beds had been described three years earlier by Jim Moore of the USGS and I had come to confirm and record his observations with photographs. Five years of erosion had created many gullies that revealed sections of deposits unavailable for observation in previous years.

The exposures of dune forms and cross bedding were superb (Fig. 8). I created a large album of photographs that illustrated the features of the dunes made by base surge beds, and later I collaborated with Aaron Waters on several scientific papers about base surge structures seen in the deposits of maar volcanoes. This research was complemented by discovery and portrayal of cross-bedded dune forms at Christmas Lake Valley in central Oregon by Grant Heiken for his Ph.D. dissertation and at Ubehebe Crater in Death Valley, California, by Bruce Crowe, another student of mine. Both men showed evidence that directions of the currents that formed the dunes, as shown by cross beds, were always directed outward from the crater, proof that the winds were not formed by atmospheric wind currents.

Suggested Reading

Francis, Peter. *Volcanoes: A Planetary Perspective.* Oxford: Oxford University Press, 1993.

Moore, J. G., K. Nakamura, and A. Alcaraz. The 1965 eruption of Taal Volcano. *Science* 151 (1966): 955–960.

Chapter Seven

◇◇◇◇◇◇◇◇◇◇◇◇◇◇◇◇◇◇◇◇◇◇◇◇◇◇◇◇◇◇◇

Mount Pelée, Martinique

At breakfast in l'Europe Hotel in Fort de France, on the island of Martinique, the French woman serving me poured only a quarter cup of coffee. "Please fill it," I asked. She refused, and I did not know enough French to coax her. When I tasted the coffee, I realized why the waitress had refused my request. It was espresso, but because I liked it, I returned for more. The second time, she filled the cup. After doing so much fieldwork with cold coffee often as my only companion, I liked it strong!

The night before, May 7, 1977, we had arrived in Fort de France, the capital of Martinique, and checked into the hotel at about ten o'clock. We walked up three rickety flights of steps, with bare lightbulbs hanging from the ceilings, to a small room that had a tiny window balcony guarded by iron filigree and overlooking a side street. The wooden floor was uneven, and it creaked like the sway-backed bed. I thought, "fire trap," and barely slept. The city of Fort de France is, of course, very French. This was my wife Beverly's and my first adventure together in a foreign country. Our children no longer lived at home.

Martinique is a department of France, equivalent to a state in the United States. The island, discovered by Columbus in 1502 on his fourth voyage to the Americas, is now a beautiful French vacation spot. On May 8, 1902, the island suffered the largest volcanic catastrophe of the twentieth century when Mount Pelée erupted and destroyed the city of St. Pierre, killing all but one of the estimated twenty-nine thousand people in the city—the survivor was being held in a basement jail cell.

We had arrived in Martinique on the day before the seventy-fifth anniversary of the 1902 eruption. The local newspaper printed several stories about people still living on the island who vividly remembered seeing the destruction following that horrible day. May 8, we checked into the Latitude Hotel at Le Carbet, closer to the volcano, and then drove on to visit St. Pierre. It was eerie to walk through the town on the anniversary of its destruction.

The climactic eruption in 1902 occurred at 8:05 A.M., when a pyroclastic flow of hot gases and fragments of rock moved as a turbulent cloud down the slopes of the volcano. The flow velocity was estimated at over sixty miles per hour and carried buildings and people to oblivion before spilling into the ocean along the southwestern coast at the foot of the volcano.

The fragments of rock and gases first blasted straight up as an eruption plume, to about two thousand feet higher than the crater's rim, and then collapsed back into the crater. The crater would have held all the debris that fell into it, were it not for a notch on the southwestern side of the volcano that extended from the crater's rim to its floor. The products of the eruption collapsed to the crater floor with enough force to project the mass outward through the southwest-facing notch and overwhelm the countryside in that direction.

The heavy lower part of the pyroclastic flow moved quickly down Rivière Blanche, filled the valley, and entered the ocean four miles from the crater. The lighter parts of the turbulent flow fanned out over and across the ridges at the foot of the volcano. The farther it traveled, the wider the fan became. The outer fringes of this fan engulfed St. Pierre, pushed over all of the stone buildings, and flowed across the water of the harbor to overturn boats at anchor. Some of the boats sank, and others caught fire from the heat of the swirling mass of hot gas and ash. The deaths were caused by falling buildings, flying stones, hot gases that burned the lungs or boiled their body fluids, or inhaled ash that clogged the noses and lungs.

Before the eruption, it was thought that Mount Pelée would not cause great damage to St. Pierre because there were too many ridges and valleys between the volcano and the city. That would have been a correct assessment if the danger had been only from lava flows, but geologists at that time were unaware of hot pyroclastic flows.

St. Pierre, formerly known as the "Paris of the Caribbean," has never fully recovered her previous luster. Her population in 1977 stood at about eight thousand, considerably less than the twenty-nine thousand who perished in 1902. Walking through the streets, I could see evidence of the past destruction. Most impressive to me was that present-day buildings used the partial walls left standing after the destruction seventy-five years previously.

The Perret museum contained artifacts, photographs, and maps of the destroyed city. I remember a batch of fused nails, a graphic reminder that the destroying cloud was extremely hot and had started fires. The cathedral, near the city center, has never been rebuilt and is now a scattered group of stones; one part of a wall, about six feet thick, still stands in an overturned leaning position to remind visitors of the holocaust. The greatest reminder is Mount Pelée itself, looming 4,600 feet above sea level and only five miles from the city's center (Fig. 9). I felt the menace of its massive presence and for a moment was filled with dread, not for my present safety but for the future. The mountain is still alive, for I have walked from her southern foot into the headwaters of the former, largely buried, Rivière Blanche and felt the hot waters of her geothermal springs rising from the depths.

In the early 1970s, the French government started a large project to resurface and upgrade roads on Martinique and discovered that the 1902 eruption had left millions of tons of loose, pulverized pyroclastic debris that required only a backhoe to load it into a truck and sieves to sift it into size grades ready for

Figure 9. Mount Pelée, Martinique, towering over the port town of St. Pierre five miles away. The eruption of Mount Pelée in 1902 produced nuées ardentes that followed a canyon (Rivière Blanche) down to the sea and away from the city. The top of the nuée expanded, however, fanned out over the countryside, and destroyed St. Pierre and all but one of its inhabitants. It erupted again in 1929–1930 and has been quiet since that time. The volcano continues to be monitored in case it again awakens. (Photo: Richard V. Fisher.)

application. There had been a similar eruption of Mount Pelée in 1929, but the pyroclastic flows were much smaller and the distribution of materials was not extensive. The French opened dozens of quarries at the foot of Mount Pelée, along four to five miles of coast between St. Pierre and St. Philomène. None of the outcrops in the quarries in which I worked in 1977 have survived; in fact, nearly all of the deposits of the pyroclastic flow have been removed for the highway project. I had come to do the study at just the right time—after many quarries were opened and before the deposit was recycled into roads.

The 1902 deposits at St. Pierre were only about a foot thick in town, but toward the northwest, the deposits were much thicker—up to twenty feet or more. I spent five days describing the rocks and photographing distinctive layering, such as dune-shaped features and coarse-grained rubble, in a large active quarry at a place called Canonville, which at one time was part of a large plantation and a small village.

My preliminary results at the quarry showed several layers of deposits: a thick lower part deposited by the pyroclastic flow overlaid by thinner deposits with dune shapes and internal thin layers that indicate deposition by wind—in this case, volcanic wind of the pyroclastic flow. This confirmed the eminent French geologist Lacroix's description of the live pyroclastic flows that came down the mountain. He recognized that the flows, which he named *nuée ardentes,* consisted of two main parts: a ground-hugging "glowing avalanche" of particles and hot gas and an overriding turbulent glowing cloud. The high temperatures of the gases, caused the basal part of the flow and the upper cloud to glow at night as they moved.

During those ten days in May 1977, I became familiar with many localities of the 1902 deposits and somewhat familiar with different parts of the island, but my most concentrated time was at the large quarry at Canonville. On the first day, I soon became aware that the area was occupied by a plantation before the eruption, for I discovered small pieces of ceramic dishes of blue willowware design, parts of perfume bottles, and other broken bits of glass. I found a small hand of a child's doll. The quarry operators had uncovered a house that had been crushed by the weight of ten feet of ash. The roof's burnt timbers were smashed flat against the tile floor, which sat on pre-1902 volcanic soil, so the house had certainly been crushed and buried by products of the 1902 eruption. Before coming to Martinique, I had read about the eruption and its impact on the budding science of volcanology, but I had not considered the human side. Seeing the artifacts and the crushed house was a genuine lesson about the human tragedy, and I confess that, for the first time, I considered that what I discover might help other human beings in the future.

This first Martinique research effort was financed by a small grant from the University of California, and I could stay for only ten days. But I had sufficient preliminary data to show the necessity of the research and to present a well-reasoned request for funds from the National Science Foundation. One of the reasons was that the nuées ardentes at Mount Pelée had been initially used as the type examples to interpret deposits around the world as nuées ardentes (and later, pyroclastic flows), yet no one had *ever* sufficiently described the deposits at Mount Pelée.

My National Science Foundation grant proposal was accepted and funded, and I returned in June and July of 1979 to complete my study. Beverly and I made arrangements to stay at the Mont Auberge, an inn with several bungalows that sits on the north side of Mount Pelée near the trailhead to the top of the volcano. We had eaten at the inn one foggy night in 1977 and vowed that when I came back to Martinique to finish my research, we would stay there.

I had returned to finish describing the characteristic features of the *Peléean* pyroclastic flows and to collect more samples to determine the physical properties of the grain fragments. We made friends with the owners of Mont Auberge as best we could with our limited French. The husband was a native of Martinique, and his wife was a native of Paris. In a halting French conversation one evening, I understood that the husband wanted me to take him into the field and show him what I did.

I said, "OK, but the quarries are barren places, and I spend hours in one place writing my descriptions of the rock layers." I thought he said that it would be all right and we could "go the next Sunday."

The next Sunday at 6 A.M. there was a loud knocking at the door and there he stood with his son; he made me understand that it was time to go. It was raining, he had on a T-shirt and shorts and was barefooted. We often saw him barefooted, but I did not know then that he had never worn shoes in his life. His son wore a raincoat and shoes, and I detected a Paris influence.

We drove in the rain up the mountain to the end of the road and the trailhead, got out of the car, and began walking up the trail that led to the top of Mount Pelée. By this time, I was finally aware that *he* had invited *me* to go with him to the top of the mountain—he had not invited himself to join me for a day in the field. It continued to rain as we climbed the steep trail, and within about fifteen minutes we entered the clouds that shrouded Mount Pelée for most of the summer that we were there. For the rest of the morning, I followed the barefoot Martinique guide and his young son through the rain and fog. We eventually came to a trail that went steeply downward from a ridge that was the rim of the 1902 volcano, dropped down into a moatlike gully, and started up again onto a *dome* that filled the old crater—still in the fog, still in the rain. I could see nothing but the lava rocks next to my feet, and I had no way to locate myself on the map, but he had shown me the top of Mount Pelée and seemed pleased. He had come the entire distance without hesitation in his bare feet, in places over sharp rocks. About noon, we began retracing our steps and two hours later, we were back at the car, soaking wet and cold. I was bone tired. Now I had seen the destroyer up close, shrouded in mist; before, I had only known her from the quarries at her feet. I knew what she could do, but didn't know why. No one knows why.

Lacroix had observed that nuées ardentes were capable of separating into two parts, and my work on the deposits showed that each part could flow independently, each capable of horrendous destruction. This was later confirmed by studies of the 1980 Mount St. Helens eruption (chap. 10).

At the end of my 1979 field season in Martinique, instead of heading straight home, we packed and headed for St. Vincent so that I could see new pyroclastic flow deposits up close. I had no idea that, instead, I would visit a gate to the underworld.

Suggested Reading

Thomas, G., and M. M. Witts. *The Day the World Ended*. New York: Stein and Day, 1969.

Chapter Eight

◇◇◇◇◇◇◇◇◇◇◇◇◇◇◇◇◇◇◇◇◇◇◇◇◇◇◇◇◇◇◇◇◇◇◇

Soufrière, St. Vincent

"Good god, it's hot." The stewardess complained in a British accent. She was tall and lanky, and in the tropical humidity, sweat poured from her body as her dark hair matted against her forehead and temples. Dressed in jeans and a baggy T-shirt that had a rip in the sleeve, she was throwing luggage into the netting at the back of the plane behind the passenger seats. I remarked to Beverly the obvious: "She's not French, she must be from St. Vincent." It was July 6, 1979, as we boarded an old World War II propeller-driven DC-3 owned by Air Martinique: we were on our way from Martinique to St. Vincent.

I had spent more than a month in Martinique working on the 1902 deposits from the eruption of Mount Pelée, but Soufrière Volcano is only ninety miles from Martinique, so I decided to compare its new nuées ardentes deposits with those from the 1902 Mount Pelée eruption. Soufrière had erupted April 13–22, 1979. We arrived in Kingstown the capital of St. Vincent, on July 6 in the afternoon.

When we rented a car, I first fully realized that St. Vincent was still owned by the British. Driving was on the left side of the roads, most of which had deep gutters on either side to carry away the waters from frequent heavy rains.

We were shocked at how poor it was compared to Martinique. "Hey you, give me a ride!" We heard this time and again as we

Figure 10. Soufrière, St. Vincent, and the author. The day before I partic-
ipated in an expedition to the top of La Soufrière, my picture was
taken with the volcano as steam poured out of its top. I went to the volcano to examine the nuée ardente deposits on its slopes, but the crater
beckoned, and I walked down to its floor to see an active dome.
(Photo: Richard V. Fisher.)

drove along the west coast of the island to visit the volcano observatory. The road was narrow, winding, and steep. Cars were
few, and most people walked along the mountain roads to go
shopping or to work. No paved road circles the island, and we
drove to the end of the pavement to view Soufrière Volcano with
its wisps of vapors that rose above the crater rim. The volcano
was ominous with its massive presence rising three thousand
feet above us (Fig. 10). It is not generally known that Soufrière
had erupted in 1902 on the day before Mount Pelée erupted. It
had produced nuées ardentes that killed two thousand people,
and the place where we had stopped was in the pathway of one
of the killer flows. I did not attempt to study the old deposits

because I did not have the time to trace them through the jungle and dig for outcrops through the soils that had developed since 1902.

Following the 1902 eruption, the Soufrière crater, with a diameter of about one mile, partly filled with water to form a lake. For seventy years, it had been a place of recreation for the islanders, but from early October 1971 to late March 1972, Soufrière erupted again, not explosively but enough to spark a hotly contested debate about whether or not to evacuate the citizens that lived near the volcano. A large hot lava dome rose steaming and hissing to form a small island in the lake, but it finally subsided and remained quiet until 1979.

Between April 13 and 22, 1979, there had been eleven highly explosive events at Soufrière. These events produced heavy falls of ash, but the most spectacular event was the eruptive episode that blew out the full contents of Soufrière's crater: the entire dome and millions of gallons of lake water. It was all expelled into roaring volcanic plumes that jetted skyward thousands of feet.

Keith Rowley, who had been studying Soufrière, was on a flight on April 13, 1979 to check the crater. I met him a few years later at a meeting in Martinique. He related, "I had just come from Trinidad and was planning to fly over the volcano to check on possible changes in the crater. Five minutes before we would have been above the crater, the volcano erupted, sending a huge black plume into the sky." He was philosophical about it. With a wide smile, he said, "My time is not yet up."

On July 10, I had the opportunity to climb Soufrière volcano. The explosive phase had largely stopped except for occasional small explosions from the dome within the crater. Henson Almorales, volcanologist technician, and his assistant Raymond Yip Choy, of the University of the West Indies in Trinidad, were leading a group of workers from the volcano observatory at Belmont into the volcano's crater to measure the growth of the dome, and I was able to tag along. The dome had continued to grow slowly after the main explosive eruptions, and there had been several expeditions into the crater through May, June, and July to study its growth rate and composition. Determining the growth rate was accomplished by descending into the crater and measuring distances from the edge of the dome to the crater walls along seven lines.

We left Belmont Observatory, six miles south of the volcano, early in the morning in four-wheel-drive vehicles, drove to the end of the paved road on the west side of the island, then on the beach to the Wallibou River, and up the rocky river bank a short distance to the trailhead. There was little talking, and I could smell the moist air that spilled from the tropical green forest. At times there was the smell of mangoes, a source of free food to travelers through the forest.

We slowly followed the narrow trail through the forest, and I kept a lookout for real and imagined snakes but saw none. On our three-thousand-foot climb to the rim of Soufrière, we walked through different zones of damage to the forest, which were so subtle that I had to consciously try to recognize them on the way down at the end of the day. First we walked through undamaged verdant tropical forest but soon encountered abundant gray ash on the ground and on the foliage. Then came a wide zone where the leaves of the trees were brown, followed by a zone of denuded trees that were sandblasted on the side facing the crater. The damage to the trees increased as we got closer to the volcano's rim, until the trees disappeared completely.

The stripped trees had many broken branches and were uncharred or slightly charred within a zone that was up to a half a mile wide. Although only two months had passed since the explosive destruction, green buds had sprouted, some on trees that were mere stumps. Within three hundred feet of the crater rim, however, the landscape was completely barren—totally scoured of all living things. Of considerable importance to interpreting the mechanism of nuées ardentes was seeing that the devastation zone extended farther down the valleys than on the ridges. This feature alone indicates that hurricane-force volcanic clouds had flowed down the slope as a heavier-than-air-density current. The trees that were partly charred sandblasted on only one side confirmed that the eruption cloud was composed of particles and hot gas that moved along the ground at a high rate of speed. I dug a hole next to a tree and found only one foot of debris left by the nuée ardente. Because the stripped tree was at least forty feet high, it meant that the nuée ardente was dilute and did not carry a high volume of solid fragments as it passed the tree, but it was nonetheless very powerful. The climb up the volcano was so quiet and the land so peaceful, it was difficult to imagine the violence that had so changed the landscape.

The view from the rim of Soufrière Volcano was breathtaking both toward the sea and into the crater, but my focus was on the crater. I followed the rim halfway around the volcano, awe-struck at the evidence of violence and hoping the inactivity of the volcano would last out the day. The dome rose about three hundred feet above the crater floor (Fig. 11). It had an uneven surface and a small crater at its top through which beautiful blue vapors of poisonous sulfur dioxide wafted upward toward the sky in lazy curls. The quiet beauty of the sulfur dioxide escaping from the dome, however, signaled that a deadly explosive erup-tion could recur at any time, for it meant that there was a direct pipeline from the hot magma to the surface. *Seismographs* had been checked that morning to find little earthquake activity be-fore we left, and the volcano had been quiet for several days, so the volcanologists felt that it was reasonably safe to go into the crater. Signs were plain that the eruption, nonexplosive for the moment, was still continuing, for we could hear rocks tumbling down the slopes of the dome, meaning that it was expanding, and there were occasional small explosions in the area where the sulfur dioxide vapors were being emitted at the top of the dome. The eruption definitely had not ended, but it was in a slow phase of extruding lava to build the dome. The dome ceased measurable growth in October 1979.

The scene was primeval. The smell of sulfur dioxide from the rim was strong in the air. New land was slowly being added to

Figure 11. Soufrière crater, St. Vincent, seen from its rim. I walked around the rim, examined nuée ardente deposits, and took photographs of the steaming dome. Noxious blue sulfur-dioxide vapors emanated from a small crater in the dome, reminding me that the volcano had erupted only two months previously and could erupt again at any time. This photograph is a view of the dome and the crater floor with its small lake, five hundred feet below. (Photo: Richard V. Fisher.)

the earth's surface with the growth of the dome, and around the fringes of its base, steam was blasting out of dozens of holes.

As the others ate lunch, I was digging holes in the ash, noting the features of the deposits and finding similarities to the nuée deposits at Mount Pelée, thereby confirming that they had spilled over the rim and swept down the slopes.

After lunch, we descended five hundred feet to the crater floor on a crude trail, soft underfoot, that had been established during previous expeditions. In the steepest areas, we were assisted by ropes. As we entered the crater, there was a gradual increase in the cacophony from the growing behemoth surrounded by hundreds of active, whistling, and hissing steam vents. The noises grew louder and more fearsome, and the dome towered above our heads as we stood on the crater floor. I had some doubts about my decision to climb the mountain and walk into the crater while the monster was still breathing. I began to see why early Christians thought that active craters were gateways to hell. I preferred working on deposits out of the crater.

While the men were doing their measurements and establishing flag lines by which to measure the advance of the dome front, I was left by myself to explore the *fumaroles* (steam vents) and the processes by which the dome was advancing. The noise was deafening and the swirling steam, whipped by an occasional wind flurry, temporarily blocked the sight and the sound of tumbling rocks (Fig. 12). The smell was like that of boiling pasta, much different than the smell of sulfur dioxide emanating from the top of the dome. The rapid escape of steam produced many pitches of sound like wind instruments. Some roared, and some whistled, the tones depending on the diameter and shape of the vent. Many opened out as inverted cones in the shape of a trumpet. The most dangerous vents were those that were too hot for steam to condense, which left the rising gases invisible. My main fear approaching the vents was of breaking through a thin crust into a subterranean steam passage, which would have meant certain death.

The origin of the steam fumaroles was easy to explain. The rising magma of the dome had punched through the postexplosion crater floor. The surface of the floor consisted of fine ash continually wetted by rain storms. As the dome rose, it lifted the adjacent crater floor, which tilted upward toward the edge of the dome and formed a small moat within the crater floor around the dome. Rain that fell into the crater pooled in the shallow irregular moat and soaked into the ground. The water trickled downward through the crater floor sediments to encounter hot

Figure 12. Steam vents in the crater of Soufrière, St. Vincent. Steam vents (fumaroles) surrounded me. Water filtered through the porous ground with each rain. The water traveled downward into the throat of the volcano to touch the cooling magma and rose again as steam that cored its way upward, creating vents to the surface. (Photo: Richard V. Fisher.)

magma, a few dozens or hundreds of feet below. The water flashed to steam and returned to the surface, the force of the pressure forming pipes that released the steam. Standing water in the crater was a few dozen yards away from the front of the dome, and there were only two small lakes in the crater, one on the west side, the other on the east.

As I walked around the base of the dome, I marveled at the barren crater walls across the moat, for I could see a cross-section of the volcano. One layer was made of rubble, another of cooled lava, and yet another of ash. One lens of lava was clearly the cross-section of a filled gully where lava had flowed toward the sea before it congealed. I was looking at how the volcano was constructed as viewed from the inside: alternating layers formed by explosive eruptions and quiet lava flows. Some of the layers were probably made from an ancient dome that grew above the crater rim and had crumbled and avalanched down the volcano's sides.

The active dome had an abrupt front that was made of rubble—large blocks that had cooled and rolled down the front, which had a slope of about thirty-five degrees. All along my walk there were continual cracking sounds, the ping of small blocks, and the crashing of large boulders. The dome was telling me that the eruption was continuing, although at a very slow pace. The rocks were pushed from underneath by the upward-moving magma, so that the dome was swelling internally and cracking its outer surface. The outward push of the dome dislodged large angular rocks that tumbled down the steep slope. Occasionally an extra-large boulder with plenty of momentum would roll out onto the fine-grained floor of the crater, but no more than about twenty or thirty feet from the base of the dome. It was easy to surmise that one of the main ways the dome grew was by what I was observing—interior expansion and the rolling of boulders down its front.

Near the small western pond, I had leaned against a large boulder about twenty feet from the front of the base as I sat on the ashy crater floor to eat lunch. With one eye on the lookout for occasional tumbling blocks, I felt drowsy in the warm sun as I looked at the vertical, barren crater walls. A small bird flew across my vision and landed near the dome at the edge of the lake to drink. "You look lost, little bird," I said, for the environment was exceedingly hostile and I felt lost in it.

I then noticed that the crater floor next to the dome had been wrinkled into folds like rugs, and cracks in the sediment radiated outward from the base of the dome. There were small slabs of claylike layers broken and tilted upward, sloping away from the dome. I was ecstatic because I had never read a description of such features. It was proof that the dome was advancing slowly outward and was acting like a bulldozer on the sediments of the crater floor. It was now obvious that there were two mechanisms by which the dome grew outward: one by the tumbling and avalanching of blocks causing the dome to advance by building one layer over the other; the other by internal expansion, upward and outward. One could have deduced the process in the easy chair at home, but it was good to confirm it by simple observation. An additional process of dome growth that I discovered from reading observatory reports received a few months after leaving Soufrière was an occasional lateral explosion that would occur from time to time from the front margins of the dome. Such explosions occurred after rains had left pockets of moisture in the cracks of the dome. If the pocket of moisture was large enough and suddenly reached critical temperature, it could explode in a flash of steam. I would not have been so relaxed and complacent in my explorations that day had I known that was a possibility, even if very remote. Such an unanticipated steam explosion occurred in the dome of the Galeras volcano crater in Colombia on January 14, 1993, and killed six volcanologists participating in a workshop and three curious tourists, all of whom were in the crater.

Following the trip to Soufrière, I returned to Santa Barbara to write up the results of my research at Mount Pelée and to make arrangements for my sabbatical leave. While I was working on my report for Mount Pelée, in May 1980, Mount St. Helens (Washington) erupted. By that time, I was committed to pursue a research project on pyroclastic flows in Germany for the 1980–1981 academic year, but planned to return to Mount St. Helens to work on the witnessed pyroclastic flows following my sabbatical leave.

I first went to Germany in 1971 to collaborate in research with Hans-Ulich Schmincke on the influence of water on eruptions at Laacher See near Mainz. The first part of the next chapter returns to that year, which is followed by some adventures and results of research on pyroclastic flows at Laacher See during my sabbatical leave in Europe in 1980–81. With completion of my research on pyroclastic flows at Mount Pelée and Laacher See, I thought that I was beginning to understand some of the complexities of pyroclastic flow processes.

Suggested Reading

Fiske, R. S., and H. Sigurdsson. Soufrière volcano, St. Vincent: Observations of its 1979 eruption from the ground, aircraft, and satellites. *Science* 216 (1982): 1105–1126.

Chapter Nine

❖❖❖❖❖❖❖❖❖❖❖❖❖❖❖❖❖❖❖❖❖❖❖❖❖❖❖❖❖❖

Volcanoes in Europe

Laacher See, Germany, 1971

I joined Hans-Ulrich Schmincke, of Bochum, Germany, in Co-
logne, July 7, 1991, and we drove south to Laacher See, a volcano
near Niedermendig. We stayed in a room in an old stone man-
sion owned by an aristocratic German, F. X. Michels. The village
of Niedermendig is underlain by many tunnels, quarried in the
Middle Ages and later by miners of basalt columns and people
who made mill wheels. Mining ended about two hundred years
ago, and now beer is stored in the tunnels. Originally, there were
twenty-eight breweries. Today there is only one; its brew is
called—what else—*Vulkanbrau* (volcano brew).

It was my first visit to Europe. There I was in the land of my
ancestors, where Roman hands quarried the volcanic rocks two
thousand years ago and used them for building materials. And
every castle from medieval times through the eighteenth century
required that I photograph its existence.

Northern Europe was my ancient homeland. People on the
streets resembled members of my family. For the first time, I realized that my Southern California home had a severely truncated history that encompassed only a hundred years of domi-
nation by northern Europeans, where previously there had been
Native Americans and Spaniards, sage brush, and paths of dust
or mud. I became aware of the youthfulness of Whittier, the city

of my birth, with its wood-framed houses, new concrete buildings, and dozens of Christian churches, offshoots of the Protestant revolt started in Germany in the sixteenth century. Relatively speaking, Whittier was a pioneer town near the edge of the Pacific Ocean.

Walking in the streets of a medieval village is a curious adventure because the walls of the buildings lean inward or outward as if the carpenters had had no plumb bob to check the direction of the greatest pull of gravity. Most fascinating are the decorations on the buildings or the style of the buildings themselves; they are the same as the illustrations in the fairy tale books of my childhood.

Aaron Waters, with whom I did research on maar volcanoes (see chap. 6), visited the Eifel volcanic district, Germany, with Hans-Ulrich Schmincke in 1970. Hans had discovered base surge dune structures and cross bedding in pumice quarries at Laacher See near Mainz and had invited Aaron to view them. Hans had been Aaron's Ph.D. student at Johns Hopkins University, Maryland, and when Aaron moved to my department at the University of California, Santa Barbara, in 1963, Hans moved to Santa Barbara's geology department to finish writing his dissertation. Hans and I had many conversations about *volcaniclastic* rocks during his stay at Santa Barbara because he was also interested in the *sedimentary* aspects of clastic rocks containing volcanic particles. Hans returned to Germany, and by 1971 he was a faculty member at Bochum University in the Ruhr district near Essen. We later coauthored a book, *Pyroclastic Rocks*, published in 1984, that was well received by professional researchers and students, and is still in use.

While at Laacher See, Hans and I were invited by Professor von Engelhardt to visit the Ries crater, fifteen miles in diameter and located about fifty miles northwest of Munich. The ancient village of Nordlingen is built on a small rise within the crater, which was originally thought to have been formed by a volcanic eruption.

81

The Ries crater is an unusual earth feature, for we now know that it was formed about fourteen million years ago when a meteorite struck the earth. The momentum of the impact was so great that considerable heat was generated, and the earth behaved like a fluid at the instant of impact.

Let me explain the probable course of events of such an event. Imagine a bowl of water. Now imagine a drop of water falling into the bowl. Often, what happens occurs too fast to see, but slow the action, and upon impact of the drop with the surface of the water, a bowl-shaped depression will form before the drop disappears. The consequence of the drop, however, continues to be felt in the water, for the drop cannot compress the water beneath it. The water then rebounds and ejects a plume of water into the air, which falls back and creates another smaller rebound, and so on until the water surface returns to its quiet condition.

The earth behaves similarly but is more complex, because it quickly changes from a fluid state to a solid. The solid meteor struck the earth and formed a deep bowl-shaped crater where the rock was shattered and thrown out around the margin. I was shown a block of *limestone* one-half mile in diameter that had been thrown three miles from the crater. The rock directly beneath the striking meteor was heated and partially melted, and rock deep beneath the surface was compressed and then rebounded into the sky as a large plume of debris broken into small and large pieces. The high-flung rock-fragment plume, mixed with hot gases generated by the high-velocity impact, reached a certain height and then fell back to earth. Unlike the water droplet, the fragmental solid material could not merge with the hard rock it encountered when it reached the surface. It flowed across the land as a turbulent density current, much like a pyroclastic flow. In the basin formed by the impact, a small mound was left from the rebound. That was the rise on which Nordlingen is built.

In addition to providing a brief introduction to European culture, the lessons learned at Laacher See were intellectually very

Figure 13. Laacher See, Germany. Hans-Ulrich Schmincke and I worked together gathering data on the Laacher See base surge dunes and cross beds. Here I am in July, 1971, cleaning an outcrop for better viewing while standing on a wave crest as if surfing a surge. (Photo: Hans-Ulrich Schmincke.)

satisfying. Laacher See is a volcanic lake occupying a crater about one and a half miles in diameter that formed thirteen thousand years ago. Hans and I viewed it as a large maar volcano, because it was surrounded on all sides by deposits with large-scale sinuous layers and streamlined dune shapes extending outward beyond the rim of the crater to two or three miles (Fig. 13). The cross bedding told us that the layers had been formed by a large and turbulent current coming from the direction of the crater, and, as further evidence, the dunes decreased in length and amplitude with distance from the crater. That was definite proof that the transport medium had come directly from the crater.

Europe, 1980–1981

Hans-Ulrich Schmincke met Beverly and me at the Dusseldorf airport on the cool morning of July 1, 1980, and took us to Bochum in the Ruhr district of Germany. It was Beverly's birthday, and we celebrated it by sleeping the rest of the day on a makeshift bed consisting of two pads on the floor of the Schminckes' apartment. The Schminckes were in the process of remodeling a house they had leased near Bochum in Wittenheven and were preparing a downstairs apartment for us. We knew Hans and Irmgaard quite well because they had lived in Santa Barbara for six months in 1976 with their children, Anna, Polly, Paco, and Max while Hans and I worked together.

In 1979, I was awarded two honors: a Senior U.S. Scientist Award from the Alexander von Humboldt Foundation to study in Germany for a year with expenses paid, and a National Science Foundation Grant to study pyroclastic flow deposits in Europe. These awards enabled me to return to Laacher See and combine studies on base surges and pyroclastic flows.

It was a year of unmatched adventure for Beverly and me. My research and the cultural opportunities of Europe gave us the learning experience of a lifetime. I was able to examine ignimbrites and maar volcanoes in Germany, Italy, and Spain and to establish a detailed research regime at Laacher See, where I had worked with Hans Schmincke in 1971. It was easy to combine my scientific objectives with our cultural goals on trips to Amsterdam, Paris, the French Riviera, Florence, Rome, Naples, the Canary Islands (Spain), the tip of Sicily (with Etna volcano), and the volcanic Aeolian Islands north of Sicily, including Lipari (where I saw magnificent base surge structures) and Vulcano (the source of the word *volcano*). The famous cities provided cultural stimulation; the land between the cities provided geologic stimulation because many places in Europe are underlain by volcanic rocks. Between July 1, 1980 and July 1, 1981, our travels in Europe totaled thirty thousand miles.

We moved into our apartment near Bochum on July 28, and I began to prepare for a volcanological meeting in the Azores to be held by the International Association of Volcanology and Chemistry of the Earth's Interior (IAVCEI) August 3–10. The highlight of the meeting for me was exploring the volcanic island of San Miguel on several field trips. One of the most beautiful sights was gazing from the rim of Sete Cidades caldera into its double crater, each holding a lake surrounded by lush green forests. We examined many outcrops along the roads, viewing former channels and pumice-rich fallout layers. The rocks told us of violent explosions, pyroclastic flows, and turbulent surges that had once traveled across this now idyllic rural setting. Much of the land around the crater lakes is tilled by farmers, and adjacent to one of the lakes is a small village where we had a lunch of fried beef, eggs, and plums.

Hans Schmincke, Kazu Nakamura, Steve Self, Mike Garcia, Rodey Batiza, Shigeo Aramaki, some of my volcanologist friends, and I took several exploratory field trips to look at features on the volcanic island. One of the stops was the Lombosada Water Works, a small mineral water bottling company that was connected with a pipe to a natural spring. I was quite skeptical that natural spring water could come from the ground already carbonated with carbon dioxide gas. I reasoned that the gas would separate from the water long before it came to the surface, comparable to soda pop that sits overnight without a lid.

"They must add the carbon dioxide after the water is put into the bottle."

"No, Fisher, the spring is naturally charged."

"I don't believe it," I said and decided to check my reasoning with a little experiment.

We followed a pipe from the bottling building to the spring where water was seeping out of the bare rock and into the pipe. I bent over to taste the water as it flowed from a small hole in the ground and found that my reasoning had been faulty. The water was unquestionably carbonated and tasted exactly like the water that was being bottled. Untested ideas remain only ideas that do not necessarily express the truth.

Another part of that same day we viewed an electricity-generating geothermal plant bought from the Japanese, which was being installed over a geothermal steam vent. As I watched from a dirt road, a two-wheeled donkey cart carrying a Portuguese family passed by. For a few seconds, my field of view was the most modern plant generating power from steam juxtaposed with the most primitive form of animal-powered transportation.

After returning to Bochum, we visited the Auvergne district of central France to examine volcanoes near the city of Clermont-Ferrand. Most people are surprised to learn that there are volcanoes in central France. Situated west of Clermont-Ferrand, there is a field of cinder cones with lava flows, domes, and maar volcanoes geologically young enough to retain the volcano forms and central craters. Puy de Dome, one of the best known, rises 4,800 feet, resting on a granite plateau above the city. It is a protrusion of magma that produced small pyroclastic flows caused by cascades of hot rock and ash collapsing from the sides of the growing dome, similar to those that formed from the dome collapse at Unzen volcano, Japan in 1991 (see chap. 10). I took pictures and a sample of the cross bedding in outcrops along the road showing that nuée ardentes flowed off the dome as it rose. The road continues to the top of the dome, which gives magnificent views across the green French landscape. We learned that the Greeks had established a temple to Mercury long before the Catholics established a church on top of it, and where there are now microwave relay stations and TV antennas devoted to television and shortwave communication systems.

We continued along the chain of craters. I visited a quarry at Beaunit that has small dunes in fine-grained layers and then went on to Gour de Tazenat, a beautiful maar volcano filled with a lake. On our way south, we drove through the town of Pontgibaud, where we saw a line of cows walk up one of the main streets and enter the front door of an old but picturesque apartment building.

Through the snowy months of the rest of fall and winter, we stayed close to Bochum, and I continued with fieldwork at Laacher See. We spent ten days at Laacher See in October, staying at the Waldfrieden Hotel on the northeastern side of the crater. At the time, the Waldfrieden was not a typical German hotel; we called it "Faulty Towers," alluding to a television comedy show about a dysfunctional hotel. An early indication that it was atypical was that they welcomed Beverly as a painter and allowed her to paint in our room while I spent each day in the field working on pyroclastic flow and surge deposits in the quarries around Laacher See. The aroma of oil paint permeated the building, but the staff did not seem to mind. Though the staff was inefficient—one with a broken arm, the others failing to clean rooms and leaving piles of linen in the hallways—the cook served superb meals, so no one complained.

The quarries in which I worked are on the flanks of the volcano outside the crater. Some of the quarries are very active, and it was imperative to work rapidly where new exposures had been opened by the earthmovers because they could be cut away before there was time enough to photograph, describe the thickness and characteristics of the different layers, and collect samples. Many of the quarries, however, had been dug but not yet filled, and I could work at a more leisurely pace.

As in 1971, in the quarries, I discovered deposits of pyroclastic surges, pyroclastic flows, and pyroclastic *fallout*, but I could see them with a more practiced eye because of my work with maar volcanoes and deposits at Mount Pelée. The most important observations that I had missed in 1971 were instances in which the vertical changes and horizontal changes in layers, related in color and particle size strongly indicated that single flow events had become separated during movement into a lower, high-

87

density flow and an upper, lower-density, dilute flow. The lower part of the layer was thick, filled irregularities beneath it, and had a horizontal top. Above that was a thinner layer made of many smaller layers with wavy dunelike patterns. The evidence of the close relationship of the two different layers—with the upper one separated from the lower one—indicates that the lower, dense flow was moving more slowly than the upper, more turbulent one that whipped over its base and distributed particles as in a strong wind storm. Time and again, I saw similar paired layers in other quarries.

Research at Mount Pelée, Soufrière St. Vincent, and Laacher See and other places in Europe gave me considerable experience recognizing features of pyroclastic flow and surge deposits. I felt ready to tackle research on the pyroclastic and surge flows of the blast deposits from the 1980 eruption of Mount St. Helens when I returned home in July 1981.

Suggested Reading

Krafft, M. *Volcanoes: Fire from the Earth.* New York: Harry N. Abrams, 1993.

❖❖❖❖❖❖❖❖❖❖❖❖❖❖❖❖❖❖❖❖❖❖❖❖❖❖❖❖❖

Mount St. Helens

Four geologists and the pilot boarded a small helicopter in Vancouver, Washington, on the crisp morning of June 9, 1980, and flew north. My first ride on a helicopter was exhilarating, flying low and clearly viewing the countryside. First we saw the houses and rectangles of the streets of Vancouver, then the green rural fields and small farms to the north, and ahead loomed Mount St. Helens with green forests on its western, southern, and eastern sides. Steam was rising above the southern rim.

We flew around the western side of the volcano, and as we approached its northern side, the green trees were abruptly replace by a border of brown ones killed by the hot breath of the volcano. Then came a narrow band of lifeless trees stripped of most of their limbs, gray as if covered by dirty snow. Suddenly the land became a colorless desert. For fifteen miles north, northwest, and northeast of the volcano, there were no standing trees, and the land looked as if it had been wiped with a rag covered with gray paint. The barren area with large zones of blown-down trees was fan-shaped with an irregular border with the forest outside the devastated zone, indicating that it had been swept by a powerful hurricanelike force that moved across the ground.

The volcano had erupted three weeks previously on May 18, 1980. The observation and record of the eruption was more complete than for any other eruption in history. It had been a clear day, unusual for May in the Cascade Mountains of Washington, and there were hundreds of people ready with both movie and still cameras should an eruption occur. There was satellite coverage from outer space and a small fixed-wing plane from which geologists were taking photographs.

Thousands of people had visited the volcano during the fifty-three days from March 27, when the volcano first erupted, to the devastating eruption on May 18. Many verbal battles took place between citizens wanting to get close to the volcano and authorities trying to keep people at a safe distance behind blockades. Residents who had been evacuated and loggers who needed to work argued that they should be able to return to their property and their jobs. Over and over, authorities heard, "I have a right to be on this land. This is public property, and I'm a taxpaying citizen."

Because of the authorities' efforts, however, the death total was held to fifty-seven people. It would have been much greater, if not for the work of local, state, and federal employees who kept most people from getting too close, and the fact that Weyerhaueser loggers, who were felling trees north of the volcano where most of the devastation occurred, did not work on Sunday.

The three other geologists in the helicopter were Drs. Don Swanson and Peter Lipman of the USGS and my student, Harry Glicken. I had come to Mount St. Helens to advise Harry on a research subject for his Ph.D. dissertation and to survey the result of the eruption at first hand.

Mount St. Helens awakened on March 20, 1980, after 123 years of sleep, with an earthquake of 4.1 magnitude. Before any volcanic eruption can occur, magma must rise from the depths. As the magma makes its way upward in the conduit, it pushes apart the adjacent rocks, which split and fracture, causing the earth to quake. When the magma gets close to the surface and the pressure lessens, gases begin to separate out and precede the magma so that fumaroles that send gas fumes into the air develop above the passageway of the magma. If the separation of the gases is quick, explosions will occur before fumaroles have time to develop. At Mount St. Helens, the hot magma pushed into the superstructure of the volcano, and rainwater that normally seeps into and saturates the volcano was heated to form steam, causing minor explosions and reaming out small craters at the top of the volcano starting March 27. This condition existed at Mount St. Helens until the catastrophic eruption on May 18.

The course of events at Mount St. Helens differed from most known volcanic eruptions, affecting the decisions of the authorities during the crisis and also influencing the eruptive events. Rising, hot, pasty magma failed to follow a direct path upward within the vent to the summit. For some physical reason, such as a point of weakness within the volcano's superstructure, the magma was deflected toward the north while still inside the volcano, even though there was apparently no vent for it to follow. The mass of magma therefore began pushing against the north side of the volcano from within and caused a bulge to grow like a tumor. The bulge was detected and announced on April 17.

This turn of events caused the volcanologists at Vancouver to establish a new observation station on a ridge six miles north of the volcano to monitor the rate of growth of the growing bulge, the purpose being to warn personnel of the occurrence of possible landslides or avalanches that could be caused by oversteepening of the volcano's slope. The observation point, Coldwater II, included monitoring instruments and a mobile home in radio contact with headquarters in Vancouver.

The rate and amount of northward movement of the bulge warned the volcanologists that measurements must be made more often in order to catch visual early warning signs of an expected landslide or avalanche. It was reasoned that an avalanche could dam the north fork of the Toutle River with earth. Such natural earth dams quickly fill with water. They are made of loosely packed rock and soil and are easily broken through by the lake waters, which quickly erode a wide channel emptying the lake and causing life-endangering floods and mudflows

Figure 14. Harry Glicken at his post at Coldwater II observation point six miles north of Mount St. Helens. Harry's job was to report visual physical changes in the volcano, such as landslides and avalanches. In February 1980, we arranged to meet at Mammoth, California, on May 18 to discuss his career. David Johnston, who was Harry's supervisor, took over the duties of his post for the week. David Johnston lost his life in a pyroclastic surge during the eruption of Mount St. Helens the following morning. (Photo: May 17, 1980, Harry Glicken.)

downstream. Harry Glicken manned the observation station from May 2 to May 17, and on the evening of May 17, David Johnston voluntarily took over Harry's post for a week (Fig. 14). Harry had an appointment to meet me the next day at Mammoth, California, and David needed the solitude to work on his notes and escape the frenetic atmosphere of the Vancouver headquarters.

Before continuing, I must relate a story told by Harry Glicken. One day after the Coldwater II observation post was established and before the eruption, a man in a Weyerhaeuser pickup drove up the road and stopped at a place overlooking the volcano.

Harry walked over and greeted the man. During their conversation about the weather and the volcano, the Weyerhaueser man said, "We're planning to set up operations here to harvest the logs." Harry was dumbfounded and answered, "But you can't do that. This is the USGS's observation station. You'll obstruct our view." With irritation in his voice, the man said, "This is Weyerhaeuser property, and we are not going to let a volcano get in the way of our operations!"

The Weyerhaueser plans, however, were not realized. The magma continued pushing, and the bulge continued growing outward at over six feet a day. At 8:32 A.M. Pacific Daylight Time, Sunday, May 18, an earthquake shook the oversteepened slope, causing it to fail. The earthquake had started the sequence of events leading to the catastrophic blast eruption. Young David Johnston died in the blast. He witnessed the start of the eruption and radioed to his headquarters, "Vancouver. Vancouver. This is it."

On that morning, geologists Keith and Dorothy Stoffel, were over the volcano. They had chartered a small airplane for a Sunday outing to fly around the volcano and take pictures. Just before the avalanche started, they were in a position to see rock and ice at the top of the volcano slide into the crater. That event had been triggered by the earthquake. About fifteen seconds later, they saw the north side of the volcano quiver like jello and then begin to flow like a fluid toward the valley as the north side of the volcano collapsed to form a debris avalanche. One-fourth of the volcano collapsed, and the volcano was reduced from 9,677 to 8,365 feet high. It was the first volcanic debris avalanche of such magnitude ever witnessed by volcanologists. The movement of the avalanche can be described as a great dry flow of broken rock (Figs. 15, 16).

A few seconds later, a dark-gray cloud blasted northward from a barren scar left by the avalanche. The blast was directed straight toward the plane, but the pilot opened full throttle and maneuvered out of the path of the fast-moving *blast surge* or blast cloud. The blast cloud continued expanding and fanning

Figure 15. The north face of Mount St. Helens, Washington, and its wreckage seen in the foreground. At 8:32 A.M. on May 18, the volcano collapsed, forming an avalanche that traveled sixteen miles down the north fork of the Toutle River. Spread out at the foot of the mountain is the wreckage—hummocks consisting of large chunks of the volcano. (Photo: Richard V. Fisher.)

northward, blowing down trees as far as fifteen miles away (Fig. 17).

In less than a minute or so, two different major volcanic events had occurred, both more capable of complete destruction of life and property than any other form of volcanic disaster; first the avalanche, then the pyroclastic flow from the lateral blast. The Mount St. Helens avalanche left a distinctive hilly surface comprising a mosaic of fragments from large broken blocks to dust-sized particles (Fig. 16), which led to the recognition of other past avalanches at other large volcanoes in many parts of the world—for example, Mount Shasta, California, Popocatepetl, Mexico; and Galunggung volcano in West Java.

Figure 16. Close-up of a hummock in the avalanche area, Mount St. Helens, Washington. On July 2, when he took this photograph with my camera, Harry Glicken showed me features of many of the hummocks in the avalanche area that he was studying for his Ph.D. dissertation. The avalanche had continued down the river valley past the ridge seen in the far distant background. (Photo: Richard V. Fisher.)

The recognition of debris avalanches has led to the realization that volcanoes cannot grow to unlimited heights within the earth's gravitational field. Collapse is a normal process in the life cycle of many tall volcanoes.

Glicken had a choice of several projects to study for his dissertation: the avalanche, the blast surge, the fall of ash downwind from the volcano, mudflows, and pyroclastic flows that occurred later in the day. The eruption of May 18 was unusual and complex. I was interested in pyroclastic flows and wanted Harry to study the blast surge deposits for his Ph.D. dissertation, but that became a research target of several people from the USGS,

Figure 17. Five miles north of Mount St. Helens, Washington, on Johnston Ridge. The shredded tree trunk testifies to the power of the northward-moving blast, which followed the collapse of the north side of the volcano. The shredded splinters were bent northward by the powerful surge that carried away the tree. Field assistant Cy Field for scale, July 18, 1983. (Photo: Richard V. Fisher.)

and later I spent five summers doing research on it myself. On the day of the helicopter visit with Swanson and Lipman, I became intrigued with the pyroclastic blast flow because in places where it had crossed ridges and valleys, it spawned denser flows that followed valleys at varying angles to the direction of the blast current. I wondered how such transformations could develop.

After much deliberation, Harry chose to study the avalanche deposits, which had started on the side of the volcano and traveled down the north fork of the Toutle River for a distance of fifteen miles, the ruined mass of one-fourth of a volcano smoldering in the valley, filling it as deep as six hundred feet. He was encouraged and financed by Dr. Barry Voight of Pennsylvania State University, an expert on large landslides and avalanches. Harry found the avalanche deposit to be a jumbled, three-dimensional mosaic consisting of very large blocks in close contact with one another decreasing to smaller and smaller pieces including microscopic fragments.

I had interviewed Harry Glicken as a prospective graduate student on February 12, 1980, before Mount St. Helens began stirring to life. He came to my office and said that he wanted to study some aspect of mudflows or pyroclastic flows with me. Furthermore, he wanted to start at UCSB as a graduate student in volcanology in the fall quarter of that year. Because I was going on sabbatical leave to Germany for a year starting in July, we planned to meet again before I left to discuss his first year of courses at Santa Barbara.

Harry Glicken missed death in the Mount St. Helens eruption, but lost his life along with volcanologists Maurice and Katia Krafft and forty Japanese on June 3, 1991. Harry had a postdoctoral appointment studying debris avalanches in Japan, when Mount Unzen began erupting in Kyushu. Harry wrote letters to some of the people in the Vancouver office of the USGS, saying that the press's video coverage of the event was so vivid that he had decided not to visit Mount Unzen, and that he had a prior commitment to lead a field trip to Bandai-San. In the meantime, Harry had sent a fax to the French volcanologists, Maurice and Katia Krafft, whom I had met in Japan in 1981 and again at El Chichón volcano in 1983 (chap. 12):

May 22, 1991

Dear Maurice and Katia,

Some time ago you asked me to inform you if there was any volcanic activity in Japan worth coming for.

Anyway, there is currently eruptive activity in Kyushu at Unzen Volcano. The events started last fall with small ash eruptions, and just yesterday (21 May) there was an extrusion of a dome.

I currently have no plans to go down there (it is a long way from Tokyo and I've got too many other things to do) but this could change.

Anyway, please telephone me (I'm at the office until about 8:30 and then at home till 11 P.M.) if you're interested. I'll try to help.

Harry probably changed his mind and accompanied the French volcanologists to Unzen because of his insatiable curiosity. In addition, the Kraffts needed an interpreter, and Maurice was a very persistent and persuasive man. Harry went with them to the volcano and was there on June 3, 1991. They came to take photographs of pyroclastic flows that swept down the mountain and were accompanied by forty Japanese media persons. All of them were inside the danger zone at the foot of the volcano declared off-limits to the citizens of the city. Glicken and the Kraffts had expected the pyroclastic flows to follow a turn in the valley and move past them like previous flows had done. Suddenly, however, a larger than expected pyroclastic flow swept down the valley. Its dense lower part followed the valley like the previous flows, but the lighter weight, intensely hot, upper part of the current separated from the lower part and moved up the hill to engulf them. They stood thirty feet above the valley floor and about three hundred feet from the edge of a previous pyroclastic flow. It traveled with great speed and force and intense heat, but was quite dilute and could move uphill. It left behind only a few inches of ash.

On June 1, 1991, I led a field trip that included three postdoctoral candidates and four graduate students to Mount St. Helens to show them what I had learned about the 1980 pyroclastic surge flow and what Glicken had learned about the debris avalanche. On June 3, when we returned to Cispus, Washington, where we had rented a trailer for shelter, I had a message from the office. It said, "Dr. Fisher, call your wife about Harry Glicken."

On the day that Harry was killed, the students and I had examined the avalanche on which he had worked for his Ph.D. degree. I could scarcely believe the coincidence. It was the last of a series of events concerning Harry and myself, and though the events were not related, some would say that I had completed some kind of metaphysical circuit, starting with my first meeting with Harry in February 1980. Both Harry and David Johnston had died in their early thirties, both of them in dilute pyroclastic flows, eleven years apart—the only two American volcanologists who have died in pyroclastic flows, or for that matter, volcanic eruptions of any kind.

In the summer of 1982, I started working on the pyroclastic blast surge deposit at Mount St. Helens. After a stressful four days looking for a house to rent, two of our grandsons, Dominic and Calisto Plourde, Beverly, and I finally found one in Castle Rock. It was a three-hour drive from Castle Rock to the valley of South Coldwater Creek, but leaving at 5 A.M. allowed me to work a full day.

The drive began by winding up highway 154 through the greenery of the partially settled forest to the village of Toutle, then along miles of dusty and bumpy gravel road still under repair from the destruction caused by the May 18 mudflow that had flooded the Toutle River valley.

I left the highway on active Weyerhaueser logging roads in the backcountry through the mountain ranges. Occasionally, I would meet a logging truck and quickly dodge to the edge of the road, but many places were too narrow for a logging truck to pass my carryall. I prayed never to meet a truck on the narrow passageways. Worst of all were foggy mornings, when I became invisible to the truckers. My teeth would clench, my mouth would be dry, and my hands would tightly grip the steering wheel. Driving was dangerous, and I once missed disaster by seconds. A truck came over the crest of a hill and downthrottled as he started down the steep slope. He suddenly saw me and loudly sounded his horn. I could see him yelling and cursing.

With only a second to spare I came to a turnout, the only one along two miles of narrow road. The next day I acquired a map from the Weyerhaueser Company and located a system of abandoned logging roads into the region.

The destination each morning was a flat clearing used for parking near the new Coldwater Lake, formed when Coldwater Creek was dammed by the avalanche as it flowed down the Toutle River. The Army Corps of Engineers had built a spillway so that the lake would not break through the natural earthen dam. To reach the area of investigation (South Coldwater Creek, which flowed into Coldwater Lake), I had to wade across the spillway. There was a Legionnaire's disease scare at the time, for the virus had been found in the lakes of the disaster area (and is common in many lakes throughout the world). Legionnaire's is a flulike disease, and several people had died from it elsewhere. I later learned that the virus grows best in water heater's that are lukewarm the temperature kept in hospitals to avoid scalding accidents.

I looked strange during those early days of fieldwork. I would park at the edge of Coldwater Lake, pull on hip boots, put on my backpack, my dust mask (to protect against potential harmful effects of silica glass in the ash), and wade across the spillway with spade in hand, trying not to inhale mist from the water. I would leave my hip boots on the other side of the spillway and head for South Coldwater Creek. The walk up the wide, flat, barren valley floor was awesome. Until my field assistant arrived two weeks later, I was alone in a silent gray land without communication with the world outside, although I felt safe because I was required to telephone the forest service management team when I entered the red zone and when I left it and tell them where I would be that day. If I did not report in the evening, the police would have come to investigate.

In the field, I often wore a red shirt to color the drabness of the landscape, and each time I would be visited by a hummingbird looking for nectar in the middle of the barren desert.

Those long walks up South Coldwater Creek were exhilarating with discovery. I could see gigantic gouges where soil had been scraped off the steep canyon sides by the airborne avalanche. The soil had been completely stripped down to bedrock in many places, and I surmised that part of the avalanche had whipped like a hurricane over the top of the ridge that faced Mount St. Helens and then crossed South Coldwater Creek in the air and slammed into the opposite ridge, scraping away several feet of soil and all of the trees growing there.

The ridge between South Coldwater Creek and Mount St. Helens was later named Johnston Ridge in memory of David Johnston, who lost his life there. That area can now be reached in less than an hour from Castle Rock by a new paved road that leads from Interstate Highway 5 to the David Johnston Observatory located on Johnston Ridge.

The former valley of South Coldwater Creek was filled with six hundred feet of avalanche deposits and pyroclastic flows that had drained from the sides of the steep canyon sides. Above the blocks of the volcano that had been carried across the thousand-foot ridge by the avalanche, there were thin sequences of tuff left by the blast surges that had swept over the landscape. Later, I discovered that the lower part of the blast surge that had deposited on steep slopes had run off as dense pyroclastic flows to collect within and flow down some of the valleys. The ravaged land spoke of the vast powers of nature that no man-made dam or barrier could impede.

My days consisted of digging trenches into the ash that had been deposited by the blast surge cloud, describing the details of the deposits, searching for clues of how the particles were carried and deposited, and detailing the features of the ash so that blast surge deposits could be recognized at other volcanoes. I learned that the shearing or frictional action at the base of the blast cloud had been so great that it smeared out the underlying soil and partially mixed with it. Such a feature had never before been reported, and, if found in old deposits on other volcanoes, can be used to interpret them as blast surge deposits.

In the summers from 1982 to 1987, my studies gradually expanded to include most of the 230 square miles of the blast surge, and I had information from 450 trenches and hundreds of sketches and photographs. It was an intriguing and complex problem; the blast took no more than five minutes to pass any one locality, but it took me eight years of research to understand partially what had happened. The final technical report of my research ("Transport and deposition of a pyroclastic surge across an area of high relief: The 18 May 1980 eruption of Mount St. Helens, Washington") was published in 1990 in the Bulletin of the Geological Society of America, Volume 102.

I employed a series of assistants to accompany me into the field, help dig trenches, and carry rock and ash samples. All were students at the University of California at Santa Barbara—Art Vaughn (1982), Cy Field (1983), Greg Valentine (1985), Michael Ort (1986), and David Buesch (1987). Valentine, Ort, and Buesch earned their Ph.D. degrees under my supervision and are now professional geologists. Valentine does research at the Los Alamos National Laboratory; Ort is a professor teaching and doing research at Northern Arizona State University; and Buesch does research with the USGS.

The development of the blast deposits at Mount St. Helens was quick and complicated. The first stage was the blast surge, a turbulent gas and particle-laden hurricane that blasted straight from the volcano outward to the north, at three hundred miles an hour or more. I call it the "transport system," because it transported all of the particles to their destinations over the 230 square miles of devastation, and toppled the trees or carried them away. The constant gravitational pull on the particles caused them to become more concentrated near the base of the blast surge and to be deposited in the final stage of motion after landing on the ground (the "depositional system").

The thousand-foot ridges and valleys caused many irregularities in the distribution, thickness, and coarseness of the ash. Where the blast flow ran parallel to valleys, it traveled farther than where it encountered high ridges at right angles to the flow.

Changes in topography caused the boundary of the destruction zone to be irregular.

My research at Mount St. Helens showed that secondary pyroclastic flows formed in separate basins where the blast surge was blocked by high ridges and drained down slopes into valleys as pyroclastic flows. Most importantly, it confirmed that pyroclastic flows can separate into heavy and lighter parts that behave differently and can form separate flows similar to the nuée ardente deposits that I observed at Mount Pelée and at Laacher See. In terms of volcanic hazards, it saves lives to know whether or not people are in the killing range of a low-density turbulent flow that moves over mountains.

Suggested Reading

Perry, R. W., and M. K. Lindell. *Living with Mount St. Helens*. Pullman: Washington State University Press, 1990.

Place, M. T. *Mount St. Helens: A Sleeping Volcano Awakes*. New York: Dodd, Mead, 1981.

Chapter Eleven

◇◇◇◇◇◇◇◇◇◇◇◇◇◇◇◇◇◇◇◇◇◇◇◇◇◇◇◇◇◇◇◇◇◇◇◇

Mount Vesuvius

It was a cold winter day in January 1981, and I was in Naples on my way to the crater of Mount Vesuvius. Naples traffic was nearly gridlocked that morning—its normal condition. I yelled at no one in particular, "Hey! It's my turn," as the drivers kept driving through the red light. My light was green, but I couldn't move without being hit by another car until I found that going into the intersection would make the other drivers stop so I could inch my way through. I had the most trouble on a one-way street marked "Sensu Unico," because cars were driving the wrong way. I could only dodge them.

At certain vantage points along the waterfront road, I could see the mountain, only ten miles or less from the center of Naples (Fig. 18). Several villages that lie at the foot of the volcano and part way up its side merge with the city. Naples and its environs are in a vulnerable position, well within the reach of pyroclastic flows that Vesuvius could send down its slopes, although, in the past two thousand years, none have reached as far as the present-day suburbs.

Eventually I found my way to a parking lot on the side of Mount Vesuvius. The road up the volcano had many views of the Bay of Naples and its coastline, especially clear during the winter months. A rocky trail up the side of the volcano started at the edge of the parking lot. The trail was littered with pieces of volcanic bombs, *blocks,* and small landslides of loose *scoria.*

104

Figure 18. Vesuvius of song and legend. Vesuvius rises above the coastal plain as modern-day farmers till its fertile soil higher and higher up its sides. The slopes of the volcano sweep down to the city of Naples, whose future is closely linked to the volcano's behavior. (Photo: Richard V. Fisher.)

I walked slowly up the trail and soon caught up with an incredibly ill-prepared group of tourists. Most of the women wore high heels and dresses; the men wore coats, ties, and dress shoes. Not a single person carried water. No one had adequate cover from the cold. No one had a backpack. But everyone wanted to see the crater of Mount Vesuvius.

As I passed each person I said "hello," but each stared at me as if I were intruding, until I finally happened upon a young man who returned my greeting.

"My name is Richard. Where are you from?" I asked.

"From the ship down in the bay. You can see it from here," and he pointed it out to me. "My name is Ivan." He spoke English well but with a heavy accent that I couldn't identify. He caught himself from falling as he stepped on a loose rock. I

105

looked quizzically at his suit and street shoes. He responded to my unstated question.

"My group is from Moscow, and we only have two days in the harbor, so a few of us decided to climb the volcano. Mount Vesuvius is very famous. I am uncomfortable, but ours is a luxury ship and no one brought walking clothes."

I said I was an American, which elicited a frown, and he said candidly, "I don't know if I should be talking to you. My government is the enemy of your government."

"But I'm not the government," I said, "so I'm not your enemy."

He liked that answer and smiled broadly. "How true," he said, "I've never met an American before. You look like an ordinary Russian from Moscow."

I took that to mean that I didn't look like a monster. "I am glad to have met you, good-bye," I said, and he said good-bye as I walked on ahead of the group.

I was within sight of the crater rim and just about to walk onto it when I encountered a man in a booth. He pointed at a sign that said in three languages, "Entry fee, two thousand lira." He repeated it in Italian. I was astonished. Entry fee for a volcano? I thought, this must be the only active volcano in the world that has a last-minute fee charged near the rim. The man smiled proudly, "Coca-Cola?" he asked. At his stand, I could rent walking shoes and buy postcards and books about Vesuvius, Pompeii, and Herculaneum, videotapes, souvenir dolls, and decorated ashtrays. I paid the fee, walked a few feet, and there I was on the rim of famous Vesuvius, looking down into its crater. The crater had steeply sloping sides shaped like a funnel that narrowed to a small irregular floor of rock about eight hundred feet below the rim. It looked like a very large cinder cone that had lava flows emanating from its outer flanks. I was surprised to see steam being emitted from several places within the crater, but I should have expected it because the last eruption took place less than forty years before.

The present cone of Vesuvius started growing about seventeen thousand years ago inside an older and larger crater called Mount Somma. It has not stopped, and the present elevation of Mount Vesuvius is 4,200 feet above sea level. There is a crude trail around the rim, which I followed. By the time I was halfway around, I could see the Russian party beginning to walk along the trail, some stumbling and some walking off the trail toward the crater. I later wondered if they all returned safely.

The view from the rim was spectacular. It was inspiring, and I was on Vesuvius! I had become a tiny part of the seventeen thousand-year history of this volcano. To the south were the excavations of Pompeii, so vulnerable in A.D. 79 on the flatlands at the foot of the volcano. To the west was the stark blue Bay of Naples with the metropolis of Naples on its north, and Sorrento Peninsula on its south; still farther west, I imagined that I could see the Isle of Capri. Eastward lay the Apennine Mountains, tall and rugged peaks six thousand feet or more in height, hazy blue in the distance.

Though much of the view was obscured by the Somma, around the base of the volcano were flat plains of rich soil, densely populated with farms growing cherries, olives, lemons, oranges, grapes, almonds, hazelnuts, and tobacco.

For thousands of years, explosive volcanic eruptions, have blanketed this region with volcanic ash made predominantly of glass particles. From nature's magic of chemical interactions, glass is easily decomposed by the weather, releasing many nutrients necessary for plant growth—potassium, calcium, sodium, trace elements, and many other minerals. Furthermore, it chemically transforms easily to various kinds of clay that form a framework to hold water, exchange nutrients, and establish a firm foundation for roots to hold the plants upright. The process is the gift of life from earth to its inhabitants.

Later that day, Beverly and I, and our friend June Ruiz who was visiting from Santa Barbara, drove to Pompeii. We were terribly disappointed because the staff that kept the ruins open to the public was on strike, and no visitors were allowed on the grounds of ancient Pompeii. We ate dinner at a restaurant that evening in Sorrento and struck up a conversation with an Italian who told us that, in November 1980, the ruins of Pompeii were damaged by a big earthquake that had its source in the Apennine Mountains. The institution that managed the ruins asked the Italian government for the equivalent of three million U.S. dollars for repairs, but their request was turned down because Pompeii was already in ruins!

The history books tell us of a variety of eruptions of Vesuvius through the centuries. The rocks of the many quarries around the volcano, tell the same story. Some eruptions have been highly explosive (*Plinian* type), producing deposits of pyroclastic flows and falls of ash without lava flows. At other times, ash has been transported into the sky by vertical eruption columns (Strombolian type) to fall back to earth as scoria and ash layers accompanied by lava flows that poured from the crater or vents on the flanks of the volcano.

Over the two thousand years since the destruction of Pompeii, it has been discovered that the longer the volcano remains inactive, the more likely it is for the next eruption to be highly explosive—the longer the time between eruptions, the greater the pressure buildup. The last eruption before A.D. 79 had been eight hundred years previously. This was time enough for erosion to soften its features and for vegetation to grow in abundance. The rich volcanic soil in the Naples region attracted and supported a highly populated region, but the Vesuvius of two thousand years ago had been asleep (dormant) for so many years that most inhabitants of the region did not know that it was a volcano.

In 71 B.C., Spartacus, leader of a slave revolt who had deserted from the Roman army, and his group of seventy-four ex-slaves escaped from the gladiators' school at Capua. He and his fellow escapees used the crater of Mount Vesuvius as a fort when they were attacked by an army of three thousand men from Rome led by Claudius Glabrius. According to the story, the slaves surprised the army from behind and were victorious. Spartacus ruled the Campania region for several years after his victory. (Paliotti, 28–29)

In A.D. 19, Strabo wrote the first description of a volcano.

> It is a mountain covered with fertile soil, whose summit seems to have been truncated horizontally. This summit forms an almost completely flat plain which is totally sterile and the color of ashes, where every so often you come upon caverns filled with fissures and formed of rock blackened as though it had been exposed to fire. From this we can assume that once there was a volcano there which became extinct after consuming the flammable material which fed it. Perhaps this is the reason for the extraordinary fertility of its slopes.   (Paliotti, 26)

There were two precursor events for the eruption of A.D. 79. The first was an earthquake on February 5 in A.D. 62, which damaged structures in Pompeii, Herculaneum, and Naples; it occurred seventeen years before the eruption. Second, and coincident with the earthquake, was the development of fumaroles that poisoned several hundred sheep. These incidents may not have been immediately connected to the eruption that destroyed Pompeii, but they indicated that the volcano was still active.

According to the letters of Pliny the Younger, quoted below, the eruption of Vesuvius began on August 24, A.D. 79 at about 1 P.M.. Ash, *pumice*, and soil layers now exposed in sections of *strata* in pits, quarries, road excavations, or soil banks that have been eroded along rivers, show that white pumice began to accumulate in the area of Pompeii south of the volcano probably from the 1 P.M. eruption. Other sections show an earlier layer of pumice from an eruption that had not been witnessed; it probably took place the previous night or early in the morning of August 24. A collective record of the entire eruption was laid down within several quarries around Mount Vesuvius. No single quarry shows all of the layers from the sequence of eruptive episodes.

A Plinian eruption column takes its name from Pliny the Elder, a Roman nobleman and one of the prominent men of the day, who died during the A.D. 79 eruption of Vesuvius. The eruption was described by Pliny the Younger, his nephew. Roman historian, Caius Cornelius Tacitus asked for information about the eruption for inclusion in a history book that he was writing. It is worth repeating the words of Pliny the Younger, because he wrote the first description of a pyroclastic flow. He was eighteen years old when he observed the eruption and wrote these recollections when he was about twenty-four.

> Your request that I would send you an account of my uncle's death, in order to transmit a more exact relation of it to posterity, deserves my acknowledgment; for, if this accident shall be celebrated by your pen, the glory of it, I am well assured, will be rendered forever illustrious. And notwithstanding he perished by a misfortune, which, as it involved at the same time a more beautiful country in ruins, and destroyed so many populous cities, seems to promise him an everlasting remembrance; notwithstanding, he has himself composed many and lasting works; yet I am persuaded, the mentioning of him in your immortal writings will greatly contribute to render his name immortal. . . . It is with extreme willingness, therefore, that I execute your commands; and should indeed have claimed the task if you had not enjoined it.
>
> He was at that time with the fleet under his command at Misenum. On the twenty-four of August, about one in the afternoon, my mother desired him to observe a cloud which had appeared of a very unusual size and shape. He had just taken a turn in the sun, and after bathing himself in cold water, and making a light luncheon, gone back to his books; he immediately arose and went out upon a rising ground from whence he might get a better sight of this very uncommon appearance. A cloud, from which mountain was uncertain at this distance, was ascending, the form of this I cannot give you a more exact description of than by likening it to that of a pine tree, for it shot up to a great

height in the form of a very tall trunk, which spread itself out at the top into a sort of branches; occasioned, I imagine, either by a sudden gust of air that impelled it, the force of which decreased as it advanced upwards, or the cloud itself being pressed back again by its own weight, expanded in the manner I have mentioned; it appeared sometimes bright and sometimes dark and spotted, according as it was either more or less impregnated with earth and cinders. This phenomenon seemed to a man of such learning and research as my uncle extraordinary, and worth further looking into. (Quoted in Bullard, 1968, 138–139.)

In many quarries dug for building materials around Vesuvius, workers have sliced through pyroclastic layers like the edges of pages in a book. To study the A.D. 79 eruption, these sections of rock revealed in quarry walls were measured and described in many places around the entire volcano by Drs. Haraldur Sigurdsson and Steven Carey of Rhode Island University. Armed with the physical details of the deposits, they were able to describe the progress of the eruption and the different types of processes that occurred, which they published with Winton Cornell, then a graduate student at the University of Rhode Island, and Tullio Pescatore, professor at the University of Naples, Italy, in *National Geographic Research Journal* in 1985. They gave representative information from Terzigno quarry, Pozzelle quarry, Villa Regin, Boscoreale, Pompeii, Oplontis, and Herculaneum.

During volcanic eruptions, rarely do the products of a single eruptive pulse blanket the entire volcano, and this was confirmed by examining the different sections at Vesuvius. During the rise of a high column of ash, for example, the wind may blow strongly in a single direction, and pyroclastic flows may be pushed strongly in one direction and confined only to valleys.

The eruption of Vesuvius began when magma encountered water and resulted in a magma-water eruption that lofted a low-level eruption column of fine-grained ash into the sky. Its distribution shows that the wind was blowing due east, for none of

111

the ash is found to the west, in Naples or along the coast of the Bay of Naples. In his writings, Pliny the Younger, who observed the eruption from Miseneum, writes of a high eruption column that occurred about 1 P.M. This is referred to as the beginning of the Plinian eruption column that later reached twenty miles into the sky. The high eruption column was carried by the winds to the southeast, with coarse ash falling close to the volcano and fine ash farther away. White pumice rained from the eruption plume for the first seven hours, followed by gray ash. Because the white pumice was erupted first, it came from magma at the top of the magma chamber. The gray pumice that followed came from magma farther down in the chamber.

There were nine pyroclastic fallout layers, seven pyroclastic surge layers, and six pyroclastic flow layers produced by the A.D. 79 eruption of Vesuvius. The ash that fell on the countryside collapsed rooftops under its weight. The pyroclastic surges swept down the mountain slopes and blew away houses, killing people who had not yet evacuated. On the western side of the volcano, a pyroclastic flow completely buried Herculaneum under one hundred feet of ash. The remains of people killed while running were found lying on top pumice-fall layers and buried beneath pyroclastic surge deposits.

In 1984 I visited Drs. Haraldur Sigurdsson and Steve Carey for part of a week while they were conducting research on the A.D. 79 deposits at Mount Vesuvius. I particularly remember the active quarries at Terzigno and Pozzelle on the eastern to southeastern side of the volcano. Since A.D. 79, thirty feet of pyroclastic material and lava flow have been deposited—an astonishingly rapid geologic rate of burial—indicating that the volcano is alive and still growing at a significant rate, even as measured by the life span of human beings.

Large wine storage bottles from ruined villas show that the land around the volcano was covered by vineyards owned by wine-producing estates. The excavated villa at Terzigno showed the remarkable power of pyroclastic surges. Within outcrops of thick pumice deposits are inconspicuous layers of pumice about a foot thick that contain more fine-grained ash and are a little darker than the pumice layers in which they occur. They were deposited by pyroclastic surges as indicated by their relationship to former obstructions. I traced one of the thin layers to a former building. Upon contact with the building, the layer quadrupled in thickness to enfold bricks and large chunks of former walls and roof. The layer became a thick jumble of coarse fragments of the wrecked villa in a matrix of pumice and ash. This tells the story of a building destroyed by a hurricanelike surge that blew down and incorporated a house of bricks.

After deposition of the first layer on August 24, a continuous eruption deposited nearly two feet of pumice at Terzigno, but at Pozzelle, only a mile to the west-southwest of Terzigno, thicknesses are greater. Changes in thickness of deposits from place to place are common. The presence or absence of layers and their thicknesses around a volcano depends a great deal on wind direction and the directions of movement of pyroclastic flows. Pyroclastic flows tend to be strongly oriented in one direction because of a directed explosion or the influence of valleys and ridges; thus, one village may be spared and another totally destroyed.

At one cliff, cut into the A.D. 79 ash by quarry operations at Terzigno, the lower layers formed a wavy pattern on the soil beneath the ash. "What do you think caused the wavy pattern," I asked. Haraldur Sigurdsson, with a more experienced eye around the excavations, answered, "Those were furrow rows between ridges where the grapevines were planted." That was one of the most fascinating features I had ever seen in a sequence of ash layers (Fig. 19). An average of fifty volcanic eruptions a year worldwide yields many potential archaeological sites. Although I have read numerous articles about archaeological sites preserved by volcanic ash, Mount Vesuvius is the only volcano where, I have personally observed archaeological evidence of an ancient thriving culture.

The year before I visited Sigurdsson and Carey in Naples, I went with them to southern Mexico on their expedition to El

Figure 19. Ash layers from the A.D. 79 eruption of Mount Vesuvius. The white layer and the layers above it are ash from the eruption of Mount Vesuvius that destroyed Pompeii. The white ash blanketed the preeruption landscape. The crest and trough pattern of the preeruption ground shows irrigation furrows for grapevines, which were planted on the crests between furrows before the eruption. The white stick is about three feet long. (Photo: Richard V. Fisher.)

Chichón volcano. I was working at Mount St. Helens and wanted to compare the blast surge deposits with the deposits at El Chichón.

Suggested Reading

Bullard, F. M. *Volcanoes*. Austin: University of Texas Press, 1968.

Paliotti, Vittorio. *Vesuvius—A Fiery History*. Naples: Cura E. Turismo, 1981.

Sheets, P. D., and D. K. Grayson. *Volcanic Activity and Human Ecology*. New York: Academic Press, 1979.

Sontag, Susan. *The Volcano Lover—A Romance*. New York: Farrar Straus Giroux, 1992.

Chapter Twelve

◇◇◇◇◇◇◇◇◇◇◇◇◇◇◇◇◇◇◇◇◇◇◇◇◇◇◇◇◇◇◇◇

El Chichón, Southern Mexico

It was a warm tropical evening, January 17, 1983, as I looked down from my hotel room onto the brightly lit plaza of Pichucalco, Mexico. It was filled with activity: lovers arm in arm; old women talking together; young boys and girls running and playing on the grass in the interior of the plaza; proud young parents with their babies; candy and tamale vendors; groups of men in argument. A ballet of human interaction in a mountain village in the southern Mexican state of Chiapas, near the border with Guatemala. We were there to meet the owner of a small plane who would take us to a landing at our camp-to-be, a ruined villa four miles east of El Chichón, which had erupted in March and April 1982.

We flew to our destination in two trips; the small plane could carry only four people and their gear. The first to go were Drs. Haraldur Sigurdsson and Steve Carey from the University of Rhode Island and Dr. Jose Espindola from the University of Mexico. David Gardello (a long-time friend of Sigurdsson's and camp director) and I waited for our turn by a corrugated iron garage that served as hanger for the small plane. We watched children play as their mother cleaned clothes on a washboard. The landing strip sloped down to the west and trees leaned toward it as if reaching out to recover their land, for the strip intrudes on the native jungle.

The trip took less than half an hour. The volcano was obscured by clouds, but awareness of its presence dominated the flight, for El Chichón had killed over two thousand people on April 4. From the air we could see the green jungle that suddenly gave way to a huge gray scar extending from the slopes of El Chichón. I sat in the front seat and, as we flew over the scarred land, I took out my camera and slid open the window, and the tiny plane swerved.

"Hey!!" yelled the pilot, "Warn me when you're going to do that!"

I was chastened but rewarded with fine pictures of a dune field at the foot of the volcano that we later identified as formed by pyroclastic surges. We landed on an ash-covered, bumpy dirt road and taxied up to the ruined villa blanketed with and surrounded by pale gray ash. The villa was at the toe end of the devil wind—the pyroclastic surge that had roared down the canyon from the volcano—and thus stood at the transitional margin from green jungle to barren desert.

We unrolled our sleeping bags on dusty floors inside the villa, set up the battery-run seismograph to monitor earthquakes from the volcano, and ate dinner in the patio from which we had scraped away some of the ash, beginning a week of exploration of that devastated land. Haraldur and Steve worked together; it was their project. I was a freelance observer, examining the devastation on my own to learn more about pyroclastic flow deposits to compare with the deposits of Mount St. Helens (see chap. 10).

On the first day, as Haraldur and Steve began their systematic sampling starting from the villa and continuing toward the slopes of El Chichón, I explored ahead, photographing and describing sections of the deposits. Continuing upstream, I encountered a field of dunes made of volcanic ash, which at first appeared to be wind-deposited sand dunes (Fig. 20). The low angle of the dune forms was, however, similar to dunes at Taal volcano (chap. 5). The area was once the thriving village of Vul-

Figure 20. Aftereffects of the El Chichón, Mexico, eruption. The field of ash dunes, shown in the middle ground, was formed by surges that flowed down the slopes of El Chichón on April 4, 1982. The volcano steaming in the background is about three miles from the dune field. Before the lethal surges arrived there was a thriving village known as Vulcan Chichónal built along the river. The river bed extends through center of the photograph and is visible on the right. (Photo: Richard V. Fisher.)

can Chichónal at the confluence of two small rivers, located nearly three miles from the volcano.

In the quiet of the morning, I explored the dune field, and in the greatly thinned layer between dune crests discovered the ghosts of tragedy: the leg of a porcelain doll; a piece of corrugated iron, probably part of a roof top; three broken bottles; part of a burnt picket fence bound by wire; part of a dish; a bent spoon; a burnt and curled shoe.

This had been a village with fields to tend. It was a hard life, but as in all villages, there had been love, children, and the occa-

117

sional joys and celebrations of life. The children playing and mothers laughing and cool evening meals as the sun set behind the hill that had exploded were the ghosts of my imagination. These were replaced by the silence of the morning following the eruption. The pyroclastic surge came pouring down the volcano's slope in the night to tear apart the flimsy plywood shacks, blow them down, cover them, and suffocate the sleeping villagers. Their lives were snuffed out in an instant as the scorching one-hundred-mile-per-hour wind full of rocks and ash blew across their land and left a barren field.

What we discovered is a help to volcanic hazards planning. Although the events were not witnessed, the layers revealed features that showed a succession of ash fallout, pyroclastic flows, pyroclastic surges, and their relationships on the eastern side of the volcano. Between the volcano and the former village is a ridge that diverted preeruption drainage off the volcano and toward the north, away from the village. A pyroclastic flow deposit on the western side of the ridge now occupies the northward-oriented valley. The pyroclastic flow had moved eastward down the volcano and was deflected northward by the ridge. A low-density surge at its top separated from the high-density flow and continued eastward across the ridge and swept away the village.

One incident vivid in my mind occurred in late afternoon after a day in the field, just as we were sitting down to dinner on the ash-covered front patio of the villa. Down the valley from the volcano came four people with sombreros riding donkeys. It was very picturesque. I got my camera from my backpack and was about to take the picture when the first rider swept off his sombrero in a wide arc and exclaimed in a heavy French accent, "Dr. Feeshair, I presume."

It was Maurice Krafft followed by his wife Katia and two Mexican guides. The well-known Kraffts were professional photographers of active volcanoes. I had met them at a volcano conference in Japan in 1981, and they died with Harry Glicken in 1991 (chap. 10). They were returning from two days on El Chichón taking pictures of the volcanic features of the mountain. They did not dismount, because they had a schedule to keep, and we talked for about five minutes. Maurice was a friendly and ebullient Frenchman.

"I must remind you," he said, "that you have not yet sent me the Mount Pelée references." I had completely forgotten.

"I will send them as soon as I get home," I promised and planned to write it down in my notebook.

They were on their way to catch a plane to return to France, and they soon left. I did not get my photograph of picturesque locals but once more had received a lesson in how small the world can be.

Before it erupted, El Chichón was a heavily vegetated hill with several villages on the lower slopes. With an elevation of 4,400 feet, it stood about 1,640 feet above the surrounding countryside and had a crater partly filled with water and a 1,250-foot-high dome. As indicated by radiocarbon dating, El Chichón had last erupted about six hundred years ago. Although studies after the eruption indicated that the volcano erupts about every six hundred years, neither the Mexican government nor the villagers regarded El Chichón as dangerous, even though hot springs and steaming fumaroles had emanated from the volcano for as long as they could remember. No one could remember an eruption. There were no legends about the volcano, and its slopes were heavily vegetated, showing no evidence of recent activity.

The first eruption took place 5:15 A.M. local time on March 29. The ash plume rose thousands of feet into the atmosphere, and the eruption continued for six hours. Falling rocks punctured rooftops, and ash collected in deep layers everywhere, collapsing houses. The ash cloud was driven by high-altitude winds and rained ash northeastward across southern Mexico. The eruption reamed out the center of El Chichón.

Minor steam explosions continued to widen the new El Chichón crater for five days, but on April 4, a second and third eruption occurred. These eruption columns were too heavy to rise very far and collapsed downward. The collapsing mass had great momentum and flowed down the volcano's slope as pyroclastic flows and surges. They swept down the valleys like hurricanes, carrying or burying everything in the way. The pyroclastic flow activity of April 4 killed more than two thousand people and wiped out over nine villages within a five-mile radius around the volcano. One set of surges devastated an area of about forty square miles and another devastated fifteen square miles.

On the second day of investigation, January 19, we climbed the barren slopes of the volcano to search for deposits formed by pyroclastic flows moving down the flanks and to examine the crater. For me, it was a slow climb over the rutted landscape, but the view of the crater from the rim was spectacular and worth the effort. The walls of the crater dropped precipitously from the craggy rim to a greenish-blue lake at the bottom, three or four hundred feet below. Around the rim of the lake were numerous steam vents. Most surprising to me was the narrow beach around the lake. It was bright yellow, indicating sand made of sulfur crystals. I was reminded of fanciful descriptions of hell, full of suffocating fumes of sulfur and agonizing eternal fire. Standing on the rim felt safe enough, but that night I dreamed about the unstable rim with me on it, sliding and falling into the abyss (Fig. 21).

The deposits on the rim of the crater consisted of six feet of coarse rubble and ash on top of the hard volcanic rock of the former dome. Between the two layers were vestiges of the verdant green mountain that grew there before the eruption. Roots and limbs of vegetation, largely unburned, formed a distinct dark-brown layer under the six feet of rubble from the 1982 eruption. This meant that the velocity of the ejecta from the destruction of the dome was not enough to scrape all the vegetation from the surface of the volcano at its rim. Farther down the flanks of the volcano, however, all vegetation was scraped away, indicating that the flow of broken material had picked up speed owing to the pull of gravity as it moved down the flanks.

Figure 21. The crater of El Chichón. This view into El Chichón from the crater rim shows fumaroles of steam around the margins of its lake. The water is greenish blue, and the beach around the lake is yellow from the high content of sulfur crystals. (Photo: Richard V. Fisher.)

After a week of studying pyroclastic surge deposits, it was time for me to go. On Saturday night, January 22, I packed my samples, field clothes, and other gear into my duffle bag, a suit-case, and a pilot's briefcase, and I was ready for my early-morn-ing Sunday flight back to Pichucalco. I waited until noon, but the small plane did not come. I made a quick decision to go by mule, and repacked my gear into the two suitcases, leaving be-hind my dirty clothes, rock samples, and field gear, and set out to the town of Chapultenango with David Gardello to help carry my bags and to rent mule transportation to the nearest town, which was Ixtacomatan. We found out later that the pilot who didn't show up was killed in a gun fight with bandits during the week we were at the volcano.

David and I walked for two hours along a well-traveled path through bright green jungle lightly dusted with volcanic ash. When we got to the village, the path widened into a narrow rutted dirt street with tin huts on both sides. Chickens pecked through the weeds, pigs roamed freely through the streets and yards, and a few people were walking toward a square to participate in Sunday idle talk. Because of the eruption, the village had been completely shut off from motorized traffic, and all commerce had to be moved by mule or on foot while waiting for the road to be restored.

There were small shops without signs to identify them, many huts, and an ancient church near the plaza, with a roof that had caved in long ago. Many people in their Sunday-best clothes were milling around the square, and there was a little green food stand where David and I courageously, or foolishly, gambled with Montezuma's revenge, and each ate two tacos. The town drunk was quite a nuisance, while those around laughed and tittered. Neither of us could speak Spanish, but we smiled and said "Hola" many times, as we wandered through the streets asking "Mula to Ixtacomitan? Mil pesos." The people would smile and laugh. It was Sunday and unknown to us, the walk to Ixtacomitan was to take four hours, across two mountain ridges—a sixteen-mile walk or mule ride. We got no takers for the mule until we offered "dos mil pesos." A young man beckoned for us to follow him and we waited in the front of a small shack while he disappeared into the field to get his strongest and best mule. A woman with strong Indian features served us coffee, but I well remembered drinking coffee with workers at the Volcano Observatory on Taal Island in the Philippines and contracting a terrible intestinal malady. David and I accepted the coffee from the woman. David drank his, and I surreptitiously poured mine out on the weedy lawn where we sat. After about an hour, my guide returned with a magnificent beast, tied my suitcases one to each side and asked me to get on. I did, and we started the journey at 2 P.M. One stirrup immediately broke, so I walked, finding it much more comfortable than riding. The tennis shoes that I wore were very comfortable, but they became stained red by the clayey tropical soil. I found that Chapultenango was on a high plateau; about one mile out of town, the trail started down an extremely steep incline into a valley of the Magdelena River, which was one thousand feet below.

As soon as we reached the river and crossed its bouldery waterway, we started up an equally steep incline, across huge boulders that had formed from an avalanche, but we soon found the well-trodden trail. We went up the trail and then down into another valley and then back up again through the rainforest. In places there were shacks, and we passed several people on foot carrying large loads of grain-filled sacks, packets of wood, and other things for construction and for food. It was a well-traveled trail, a necessary lifeline for subsistence in the scattered shacks and small villages along its pathway. We came to some new construction of a road bed, which to me looked like the easiest way, but we continued to take the ancient trail. Through one village, my guide requested that I ride the mule—to illustrate his guide abilities or to keep me out of harm's way—for it was Sunday, and by late afternoon, many men were drunk, drinking directly from tequila bottles while boisterously laughing in front of a small, square, tin building that served as a church, which was full of women singing religious songs, while children played basketball in the plaza. The day was beginning to darken, and it became too late to take photographs.

We finally arrived in Ixtacomatan at 7:00 P.M. It was dusk, a few street lights were on, and I could see the blue cast of televisions in darkened rooms of the village. Manuel, the guide, stopped at the pavement of the village streets and turned around for the trip back in the dark. It was two thousand pesos for the ride; I tipped Manuel one thousand pesos, a total of about twenty U.S. dollars. It had taken us five hours from Chapultenango, and Manuel had to return sixteen miles in the dark the way that we had come.

I walked up to the plaza of the village with my bags and sat down on a bench where the word "TAXI" was painted on the curb. I sat next to a man who asked me something in Spanish and I said, "no comprende." I pointed in the direction of El Chichón and said "vulcan" and pantomimed that I had been there and was returning to the United States. Before long, some more people showed up. Within fifteen minutes, I was surrounded by at least seventy-five people—so many people that mothers on the outer fringes were holding up their children to see me in the center. I was told to take the bus at 10 P.M., but the taxi finally arrived—a taxi driver and three young teenage boys who were learning English in school. They wanted to practice English.

The taxi driver looked at me carefully and said, "two thousand pesos to Villahermosa." When I nodded yes to the enormous price, the crowd collectively expressed their awe. That was a lot of pesos, but only $13.50 in U.S. dollars. I got in the taxi and went to Pichucalco, not far from Ixtacomitan. Hundreds of people were promenading in their Sunday clothes around the brightly lit plaza. The taxi driver drove me to this village to talk to the "patron," the taxi owner. I sat in the back of the taxi, and he walked up and stared at me without smiling. He was very short, wore very thick glasses, and had a butch haircut. Expressionless, he looked at me without speaking for a long and uncomfortable time. Thoughts of two recently murdered American professors went through my head. With the big disparity between the worth of dollars versus pesos, the two Americans had been killed in Mexico for their money. Finally, the patron said, "two thousand pesos." I said, "Si," and he waved his driver and his passengers to go. We hurried away into the night, leaving the village at 7:30.

We arrived in Villahermosa about 9:30 P.M. and began looking for Hotel Maya Tobasco. It took about an hour to find it. I tipped the driver a thousand pesos and, as the taxi drove away, I could hear the teenagers and driver laughing and yelling in jubilation over that enormous sum.

The total cost of the trip from the isolated villa at the foot of El Chichón volcano to Villahermosa had been about forty U.S. dollars, and it had taken ten hours of walking, riding a mule, and riding in a taxi. Ten hours from a primitive Indian village to sophisticated and large Villahermosa with a modern airport. I took a shower, went to the restaurant in the hotel, and then to bed. The next day I was at the airport at 7 A.M. and home at Santa Barbara that evening. The contrast between primitive and modern was overwhelming, and I rejoiced in the safety and warmth of home and family.

Suggested Reading

Blong, R. J. *Volcanic Hazards*. Sydney: Academic Press, 1984.

Decker, R. W., and B. B. Decker. *Volcanoes,* 3d ed. New York: W. H. Freeman, 1997.

Chapter Thirteen

Calderas

While I was in the army at Los Alamos, New Mexico, from 1946 to 1947, someone told me that there was a volcano just over the ridge that borders Los Alamos to the west, so I spent a day looking for it. I climbed to the top of the ridge, looked into the next valley, and was terribly disappointed for I saw no volcano. I was standing on the rim of the volcano looking down into the valley of the crater, but in my ignorance was irritated that I had wasted my energy looking for a volcano and found only a valley. The entire town of Los Alamos and all of the scientific laboratories were built on the outer flanks of a volcano, called a caldera, which is much too large to see from the ground.

A caldera is a type of volcano with a crater larger than five miles in diameter. The craters of calderas are formed by collapse of the ground into an underground cavity formed by the ejection of pyroclastic material equivalent in volume to that of the crater. The eruption of large volumes of material that create calderas usually produce high-temperature pyroclatic flows. The larger the volume of ejected material, the larger the caldera crater. Some calderas are as large as forty miles in diameter. Because of collapse, calderas do not rise to great heights above the surrounding landscape as do Mount Fuji, Japan, or Mount Rainier, Washington.

Many years later, remembering that experience, I led a field trip of students from Santa Barbara to the center of Long Valley, at one end of which is Lake Crowley, a fisherman's lake north of Bishop, California. "Well, here's the volcano," I said to the students. They looked in all directions across a flat, barren, grass-covered plain in the attempt to see a mountain with a crater but could not see one. They were baffled, and nearly to a person said, "Where?" I let them look and think awhile and finally said, "Right here. You are standing *in* it."

All that was visible through the shimmering heat waves rising from the ground were low cliffs ten miles away to the east with some low-standing hills to the northwest. The students had read about Long Valley Caldera before we left Santa Barbara and had seen the circular to elliptical pattern of the large crater well-marked on a map, but reading does not convey reality the same way as seeing it in person. Some calderas of the world have craters so large that they are only discernible from satellites.

Los Alamos, New Mexico is a national laboratory sited on a plateau sliced through by several canyons. The canyon walls reveal nearly horizontal layers of ignimbrite, which are deposits of pyroclastic flows that came sweeping out of Valles Caldera in two episodes, 1.2 and 0.7 million years ago. Valles Caldera is fifteen miles across. The plateau on which Los Alamos is built is part of an ignimbrite apron that forms the outer slopes completely around the caldera. Pyroclastic flows that form during caldera formation are capable of moving long distances—one hundred miles or more depending on the volume and rate of emplacement and, therefore, the outer part of such volcanoes are low-sloping plateaus.

After my experience with the ignimbrite of the John Day Formation, Oregon, in the 1960s, I became interested in calderas because they were the source of voluminous pyroclastic flows, and I wanted to know how the flows moved and deposited their materials.

I have visited only a few of the many calderas on earth. One is the Phlegrean Fields, a caldera occupied by one and a half million or more people living in and around the cities of Pozzuoli and Naples, Italy. Another, in Yellowstone National Park, is a caldera forty miles across with a gigantic volume of ignimbrite.

127

A much smaller one is Crater Lake, Oregon, only five miles in diameter. But my most memorable trip to a caldera was on the high Andes plateau of western Argentina.

Cerro Panizos, Argentina

In 1986, I made a reconnaissance trip to the high Andean caldera known as Cerro Panizos along with Michael Ort, my graduate student from UCSB; Dr. Beatriz Coria, a geologist at the University in Jujuy, western Argentina; and Dr. Mario Mazzoni, from the University of La Plata, Argentina, who had studied with me at Santa Barbara in 1981–82. We started at Jujuy, Argentina, October 10, and drove to Humahuaca where the paved road ends, and continued seventy miles on gravel road to Abra Pampa, a village on the high plateau of Alto Plano at twelve thousand feet (also known as Siberia Argentina). The driver of the car constantly chewed brown coca leaves from which cocaine is derived.

"Is the driver safe?" I asked.

Mario said, "Happy, relaxed drivers are safe drivers."

I was not comfortable because I did not believe that steady, though small, dosages of cocaine are a safe chewing habit for a driver. The driver seemed too relaxed.

In February 1986, at an international meeting of volcanologists in New Zealand, Mario Mazzoni had talked to me about a collaborative project on Cerro Panizos, a conspicuous (by satellite) but little-known source of ignimbrite in the high Andes on the Argentina-Bolivia border. He and Beatriz Coria had made a three-day reconnaissance trip into the caldera in 1985. With the Argentinian geologists' approval, Dr. Grant Heiken at Los Alamos, myself, and Michael Ort, proposed to study the caldera. We submitted the grant proposal to the Institute of Geophysics and Planetary Physics, University of California and were awarded a grant, mainly to support Ort's dissertation research.

The object of the research was Cerro Panizos caldera and its ignimbrite deposits. The satellite image of the volcano looks like a fried egg. Rather than being dome-shaped, the yolk of the egg is the down-sagged central area of the caldera fifteen and a half miles in diameter. The Panizos caldera is unique among calderas because it is filled with many small volcanoes, some of which rise three thousand feet above the caldera floor. The satellite image shows the caldera to contain several domes, like pimples or warts, within the yolk area. Comparable to the white of the egg surrounding its yolk, the ignimbrite plateau slopes away from the central caldera and has a diameter of one hundred miles including the caldera. Like Valles Caldera, New Mexico, the plateau is sliced through by many canyons whose walls are made of ignimbrite layers.

Cerro Panizos is mostly in Argentina, but a significant part straddles the Argentinian-Bolivian border. It is inaccessible by road except from the Bolivian side. From Argentina, all travel is by foot, mule, or donkey. This was Ort's field area. His purpose was to describe the structural and chemical setting of the caldera and to understand the pyroclastic flow processes involved in producing the ignimbrite plateau around the caldera. The collapse of the ground into the void after the extensive eruption of ignimbrite did not produce the cliffs formed by a ring of fractures around the crater that usually mark the edge of the collapsed caldera—a crater bounded by a cliff. Rather, where the edge of the crater ought to be, the ignimbrite layers drape across the rim; the ignimbrite slopes outward from the rim toward the surrounding valleys and inward from the rim toward the center of the caldera. This change in slope defines the location of the caldera rim that extends around the immense crater. Downward collapse into the magma chamber produced a down-sagged caldera, where the strata bent downward rather than breaking. Prior to our research, it was thought, from visual inspection of satellite images, that the caldera and its ignimbrite apron formed a shield-shaped volcano without a collapsed central part.

On our trip to Cerro Panizos, we drove along steep canyons on a good but graveled road in beautiful big-mountain country and finally arrived at Abra Pampa, a rather large, colorless Indian village along the unpaved highway. Many of the women wore brilliant colors, and I could not help concluding that the colorless environment prompts people to wear bright colors. Abra Pampa was considered a way station, where we stayed overnight to become gradually acclimatized to the high altitudes. I got little sleep, however, because I had a bad headache from the altitude. We slept on the floor of a house. My sleeping bag was next to the driver's; I could hear him chewing all night long and awakened to a disgusting, large brown pile of chewed leaves where he had spit them in the night. Chewing coca leaves is apparently quite common in the Andes—I suspect to moderate the feelings about a harsh environment.

The next day we drove mile after mile across the vast rolling desert plain of the Alto Plano of the Andes about twelve thousand feet above sea level, visiting some volcanic outcrops on the way to Cusi Cusi, the last outpost of civilization before reaching the Bolivian border. The village housed a contingent of soldiers who guarded the border thirty miles away. We stopped at one village on the way. I had some candied popcorn that I gave to a nearly toothless smiling man for his children, but he immediately began eating it and within minutes it was gone.

At dusk, we arrived at Cusi Cusi, 13,500 feet in elevation, and had dinner in the village school house—spaghetti by chef Mario Mazzoni. Cusi Cusi is at the edge of the ignimbrite plateau, which is the base of the volcano. That night we stayed in the village police station and slept on cots in our sleeping bags.

The next morning we were assigned mules. Mine was the tallest mule, and I was least able to mount him. I tried once without standing upon a rock, but the mule didn't like it and turned quickly, nearly knocking me down. I named the mule Napoleon and kept my distance when I was not mounted.

We left Cusi Cusi on October 13. Our troop consisted of two carabineros, who were the mule runners, and four geologists. We first traveled parallel to the edge of the ignimbrite plateau and then turned up a winding trail along a canyon that cut through the layers of ignimbrite.

The layers that are exposed in the canyon show a complex eruption history of at least four major episodes where several layers of ignimbrite were deposited. One of the units contains up to 50 percent crystals forming a large volume of the pyroclastic flow—more than most other known ignimbrites in the world. We collected several rock specimens from the canyon sides for chemical analysis to be done in Santa Barbara.

After about three hours, we emerged from the canyon onto a rolling plateau of ignimbrite, moving through the crisp and clear desert country at fourteen thousand feet. About ten miles from Cusi Cusi, we met a native woman carrying a baby on her back and accompanied by her young son, about seven or eight years old, and a dog. She was striding purposefully and swiftly as her boy trotted behind her. Both of them wore sandals without stockings on their feet. Dr. Beatriz Coria asked her directions to the canyon that we needed to follow into the interior of the caldera and questions about the conditions of the terrain. I later asked Beatriz, "Where did that woman come from and where she was going?"

"She came from a small village about twenty miles from Cusi Cusi to see friends," Beatriz answered.

I marveled at the stamina of the woman walking on the rolling fourteen thousand foot rocky plain. It was incredible that she could walk for twenty miles with a baby on her back, but just as incredible was her little boy trotting behind her. And I thought, "They will have to walk back!"

The vistas were broad and uncluttered by trees, the October days balmy, peaceful, and soft, but every time I managed to get Napoleon to stop so that I could take a picture, he would turn and ruin the shot. I became familiar with his moves and was able to get a few pictures from muleback.

At about fourteen thousand feet above sea level, we stopped for lunch at Abutarda Lake. I was amazed to see a flock of pink flamingos standing in shallow water, and I wondered if they had a hard time flying in the thin air at such high altitude. But there were many birds. I vividly remember flocks of small, brilliant

Figure 22. Author and Napoleon, Panizos Caldera, Argentina. The cantankerous mule Napoleon carries me toward the interior of Panizos Caldera. Within the crater area are small domes and volcanoes, seen in the background. The layer on the right side of the photograph slants toward the center of the caldera. The caldera crater has no sharp or obvious boundaries; instead, the rim is formed of layers that sag inward toward it. Toward the outer margin of the rim, the layers slant outward. (Photo: Michael Ort.)

green parakeets or parakeetlike birds. We traveled near herds of llamas and once went through a herd. They were not easily frightened and seemed annoyed when we encroached on their territory. They could easily hurt a person on foot.

We had steadily traveled up the low incline of the ignimbrite plateau but finally came to a place where we could see the trail begin to wind between small domes and volcanoes. We knew that we were in the crater (Fig. 22). The transition was barely noticeable, but we had left the ignimbrite apron and had entered the caldera itself.

We traveled until late afternoon, taking samples and photographs, when we came to a small compound with goats, an old woman caretaker, and a single-room stone house called Puesto J.C. Coria at about 14,500 feet. We camped there for the night; the soldiers slept in the rat-infested house, and we geologists slept in a tent designed to keep small (and large) animals out.

That night was long. I woke up many times unable to catch my breath. I had a mild case of "periodic breathing," which nearly always occurs at night and at elevations of more than nine thousand feet. Typically, there are four or more breaths and then a short period of no breathing for as long as fifteen seconds. The pattern can be repeated many times and often awakens the sufferer. I took a medicine called acetazolamide (Diamox) to lessen the severity of the symptoms and to prevent mountain sickness. I also awoke many times because I had to go out into the freezing night to urinate. It was difficult to move silently without disturbing the others because the tent had a tunnel opening with diaphragmlike flaps at either end to help contain warmth. It was like moving through a sausage skin and difficult to be quiet. Though I did not remain outside long, I did catch sight of the nearly full moon, which brightened the peaks around the Puesto.

The next day we continued through the caldera, reaching an elevation of about 15,500 feet, but there were peaks up to three thousand feet higher. We could not have recognized the caldera from the ground without prior knowledge. Confined within the center were many small peaks that we had observed from satellite images. At fifteen thousand feet, the land was sparsely covered with llareta, clumps of green moss. There were small streams originating from springs in which polliwogs swam and metamorphosed into frogs.

On October 15, we reached Puesto Zenon Coria at about 15,500 feet above sea level. Late that afternoon, while the others were busy with mules and tents, I sat on a rock looking at the silent and nearly barren landscape and saw several Andean condors fly overhead—and then flamingos. It was spectacular. The sightings made me once again marvel at the tenacity of life on earth. There is no environment where life does not exist—the deepest floor of the oceans, the Arctic and Antarctic, the jungles of South America and Africa, the arid deserts of the world. Here we were, close to sixteen thousand feet high, where people walk in sandals along the trails with their children, and there are flocks of flamingos, parakeets, and condors. There are polliwogs and frogs, llamas and vicuñas, flies and bees, bushes and flowers. We could not even trust the water, because it teemed with abundant microscopic life, amoebas that we filtered out to prevent serious stomach and intestinal disorders.

With the help of the old Indian woman who tended the goats, the soldiers made rice stew and roasted a goat. They made a crucifix out of two pieces of wood and tied the skinned goat to it and leaned it over a fire supported by a stone wall as it was slowly turned. The high Andes is no place for a strict vegetarian.

That day I had caught a cold. Beatriz said, "I know a mountain remedy to help you get rid of a cold. I'll make tea of coca leaves."

She boiled coca leaves in water over the fire, and I drank it. It was bland, but the hot tea felt good in my stomach on that cold starry night.

The following day we prepared to leave for Cusi Cusi. The mules were touchy and excited because they somehow understood that they were going back, although none of them had ever been to Cerros Panizos before. When we had traveled away from Cusi Cusi, each tried to lag behind. Unlike the trip into Panizos, each mule tried to be first as we headed back. For me, the trip back was very uncomfortable. By October 16, I had developed a terrible rash where I sat upon Napoleon, and when I could, I walked, holding Napoleon's reins.

Not far from the base of the ignimbrite plateau, I decided to walk the rest of the way. It was only about six miles to Cusi Cusi, and riding downhill aggravated the discomfort and burning of the rash. I dismounted and tied my backpack to Napoleon's saddle, leading him by his reins. The rest of the party soon passed us by and were out of sight, and Napoleon, very anxious, became aggressive. He walked behind but continued to push me in the back with his nose to hurry me. Several times he tried to pass, but I somehow managed to keep him in back of me. Finally, I got tired of holding him back and decided to remount and ride the last part of the trail. I tied him to a bush and was about to mount him from a rock when he suddenly tossed his head, pulled the branches from the bush with his reins, and came loose. I fell off the rock, and by the time I could look up, Napoleon and my backpack were out of sight down the trail.

It was a perfect day of cool sunshine, and I welcomed the quiet interlude as I started walking down the trail whistling, with my hands in my pockets. The peacefulness was only disturbed by concern for my backpack. It was not tied too securely to Napoleon's saddle, and it held my money, travelers checks, camera, passport, notebook, credit cards, and exposed film.

It was a lovely walk, and when I reached the bottom of the canyon and headed down the valley toward Cusi Cusi, I could see the rest of the group stopped about a mile farther down the valley. Beatriz came back to see if I were all right because Napoleon had passed them several minutes ago, galloping as fast as he could go toward Cusi Cusi. The mule tenders could not stop him nor catch him. He was running full tilt. Beatriz came back because she was the only one who could get her mule to walk in a direction away from Cusi Cusi.

She called, "Your mule passed us a while ago. Are you all right?"

"Did you see my backpack?" I asked.

Neither Beatriz nor the others had noticed a backpack on Napoleon as he streaked by. We assumed that he was heading for Cusi Cusi, but no one was certain he would stop there. When we arrived an hour later, the pack and Napoleon were there; the pack had held on by a single flimsy strap, and my valuables were in order.

To celebrate the success of our trip, that evening the women of the village cooked a big meal of rice and goat. Cusi Cusi had electricity for two hours after the sun set. The military establishment in the frontier town started the generators, the dirt road through the village was lighted, and we could listen to the radio.

We left Cusi Cusi at 9 A.M. on October 17, arrived in Abra Pampa at noon, and returned to Jujuy at 6 P.M. the same day. Michael immediately began making preparations for the trip back to Cerro Panizos to begin his field work, and I left for the United States, knowing that Michael was capable of doing excellent research on the caldera for his doctorate.

Suggested Reading

Scarth, A. *Volcanoes*. College Station: Texas A&M University Press, 1994.

Chapter Fourteen

◇◇◇◇◇◇◇◇◇◇◇◇◇◇◇◇◇◇◇◇◇◇◇◇◇◇◇◇◇◇◇◇

Communist China

Mainland China is not a country I would choose for studying large and active volcanoes because there aren't many. But in 1985, I received an invitation to lecture in the People's Republic of China and seized the opportunity because I wanted to meet Chinese geologists and see their country. I also wanted to see the sixty-million-year old ignimbrites of eastern China. They occur in a swath three hundred miles wide and one thousand miles long along the coast where the country bellies out into the China Sea. A Ph.D. geology student from China had brought me several specimens from that region two years before, saying, "You should visit that land of ignimbrite."

While in Hong Kong, just before entering China, Beverly and I had taken a ferry boat to Kowloon. I well remember sitting across from a beautiful, young Chinese woman, probably in her early twenties, who was sitting next to an old, bent, white-haired Chinese woman, probably in her eighties. The contrast was stark—young and old—the cares of the world etched on the face of one, and on the other the bright-eyed optimism of the future. This was the face of mainland China when we arrived.

We were in China only forty-four days. I lectured three weeks, and for two weeks studied the ignimbrites along the eastern coastal region, which was closed to most westerners. We also arranged trips to Xian (central China) where Emperor Shih

Huang Di's terra-cotta army is located; to Beijing (northern China), where we walked along the Great Wall; and to Guilin (southern China) along the Lijiang River through the heart of the lands of the pillarlike mountains so famous in ancient Chinese paintings. A roughly equivalent trip in mileage in the United States would include visits to the northern, central, and western United States.

In early February 1985, Dr. Qu Qin-yue, president of Nanjing University, had invited me to lecture about volcanic rocks, and offered a six-week stay in China, with expenses paid and an interpreter. Chinese geological ideas lagged behind modern advances because of the largely archaic ideas learned from the Russians and because of the Chinese cultural revolution from the mid-60s to the mid-70s, which shut down universities throughout China for ten years and nearly destroyed them permanently. The Chinese were therefore attempting to catch up with modern science. In 1984, I had coauthored a book on pyroclastic rocks with Hans-Ulrich Schmincke of Bochum University, Germany, and the Chinese were eager to learn the latest ideas on the subject.

We were greeted at the Shanghai airport by two Chinese women holding a sign that said, "Welcome Dr. Fisher." The two women were researchers in the geology department at Nanjing University, Zheu Si Zhao (Tsu Tzi Saw) and Xie Jiyu (Jean #1). They walked us through a special line without immigration inspection, and we were accorded privileges given to visiting dignitaries. Jean #1 and Tzi Saw were very friendly and companionable and both spoke excellent English.

We got our first look at China through the windows of a taxi—thousands of people, thousands of bicycles. I had expected empty streets lined with empty stores, as I heard was the case in the Soviet Union in 1985, but there were people everywhere. Our drive was a mad dash down beautiful tree-lined streets. Many parts of Shanghai through which we drove were poor, but the people were clean, neatly dressed, and appeared to be healthy. Eventually we drove through a huge gate in a high stone wall surrounding many acres of manicured garden land-scape dotted with neat, simple buildings housing the College of Foreign Language. The grounds, buildings, and walkways were sedate and peaceful, shutting out many of the city sounds.

Waking up the next morning was hard because of lingering jet lag. Tzi Saw timidly knocked on the door and said, "We must go buy dumplings for our lunch and be at the train at noon." We again struggled to get our bags into a taxi, which drove through the crowded maelstrom of Shanghai honking con-stantly. At a small market, the taxi stopped and blocked traffic for five minutes while Jean #1 rushed to buy the dumplings and returned with steaming bundles. Other drivers were furious. "Oh dear, we're going to be late. We're going to be late," la-mented Tzi Saw.

She was right, we had arrived at the station a little late and needed help with the luggage to get to the right train on time. Much to the delight of hundreds of Chinese, our bags were put on a motorized cart, and all of us sat swaying high upon the luggage as we were driven rapidly down the concourse to the train. People in the crowds pointed and laughed at the west-erners in such precarious positions holding on to the bouncing luggage.

The trip to Nanjing was like a slow-motion dream. Farmland along the endless floodplain of the Yangtse River drifted by as the coal-fired steam engine chugged along, sometimes with a shift in the wind sending smoke into the cars. We traveled six hours through the flat Chinese countryside where hundreds of workers were stooped at different tasks, harvesting fields of golden stalks that looked like wheat but (as we found out later) were actually stalks producing rape seed, the source of canola oil. Some workers cut the stalks; others bound them into small bundles, while others carried them to two-wheeled carts, which were then pulled by the carters to storage. Some of the workers hoed the furrows; others threw the grain into the air to separate the seeds from the chaff. Because it was June, the beginning of a hot season without wind, the farmers at one commune had plugged in an electric fan on a five-foot stand and threw the grain into the wind of the fan. I had never given thought to the idea that modern machines could be put to use at a one thousand-year-old practice in such a manner. Later we saw Chinese farmers rake piles of rape seed stalks onto roads so that automobiles and trucks would crush the seed pods to release the seeds, which were then raked into piles to be loaded and transported by carts to storage. I vividly remember the constant popping and snapping sounds of the seed pods as our car drove over them.

The long train trip passed dozens of country villages, and in the fields surrounding each village, hundreds of people were at work, including small children. Passing winding country roads, we saw dozens of cyclists waiting at the train barriers. Well-dressed Chinese crowded the stations waiting for their trains.

In the train we sat at a table covered by a white cloth, talking, drinking hot tea, and eating the steamed dumplings bought in Shanghai. I sat by an open window taking photo after photo of farmers, people walking along dirt roads, boats in canals, fields of workers, and children at play.

Jean #1 asked, "Why do you take so many pictures?"

"To remember China," I said.

"But they are of such ordinary subjects."

"Not in America," I said.

As the huge red sun was setting in the mist, we arrived in Nanjing. Dr. Chen Kerong and Mr. Lin Chengyi, both smiling, greeted us warmly. Dr. Chen's wide smile created deep dimples on both cheeks. As a term of endearment, we nicknamed him "dimples" but kept it to ourselves for fear of insulting him. Dr. Chen, Associate Professor of Petrology at the university, was the one who had submitted my name to the university president as a candidate for updating the sciences at Nanjing University, and it was he who organized my field trips and trips to other parts of China. He shook my hand vigorously for a long time and said "I am very glad to meet you. I am so happy to meet you. Thank you for coming to China. I am very happy."

Mr. Lin, who translated for me during my time at Nanjing and on the field trips, shook my hand and said, "I am happy to greet you. I have spent a year in the United States at the University of Wisconsin. My American name is Charlie." Later he proudly wore a T-shirt with Charlie printed on the back. At Nanjing University, he served as a Lecturer in Mineralogy.

There were two cars, one to carry the luggage and the two researchers who greeted us in Shanghai and one to carry Beverly, myself, Mr. Lin, and Dr. Chen to the university. It was a short trip, but we were introduced to the hustle and bustle of a busy well-lighted city, and were greeted by Mr. Zhou Song-shan, head of the Office of Foreign Expert Affairs. He was a government laison who officially registered us as legal visitors. We needed his approval for all trips inside China. His office was close to the "Foreign Experts Building," a plain, six-story hotel building with motellike rooms where we lived during our stay.

"This is your room," Mr. Zhou said proudly. It had twin beds, a bathroom with toilet and shower, and a small balcony that overlooked red brick university buildings, grass lawns, and tree-lined walkways. Not least, there was a small refrigerator, which cooled water as the summer grew hotter and hotter. In the early mornings, we could look out our balcony and watch the early student risers doing Tai Chi. Each morning at six o'clock we were awakened by their loud music blaring across campus.

Our meals were eaten in a university cafeteria for foreigners, an occasional Chinese dignitary, students learning Mandarin Chinese, and British and American citizens who taught English to the Chinese university students. During our stay, meals consisted variously of oily, boney fish, rice, chicken with bones splintered by a meat cleaver, vegetables floating in oil, and sugared rolls. One morning the cafeteria cooks, with proud smiles, made us an American breakfast; two eggs and potatoes. The eggs were boiled in hot oil, and they floated in the oil, like the potatoes.

Before my first lecture on June 3, I met with the University's Vice President Wong; Mr. Zhou, head of the Office of Foreign Expert Affairs; Dr. Chen; Charlie; and Dr. Zhan, chairman of the geology department. We sat in overstuffed chairs and drank hot tea served from large thermos bottles. Nearly everywhere we went in China, tea or hot water was served from thermos bottles. All drinking water was boiled.

Later in the day, came an embarrassing watermelon meeting with several geologists in the department. We sat on overstuffed sofas, and everyone was handed a slice of watermelon, for it was June and watermelon season. Everyone ate watermelon with great sucking noises without dropping a seed or spilling a drop of juice, except for me. Juice ran down my elbows and dripped onto the carpet as I leaned over to keep it from getting on the sofa or my pants. A little pile of seeds were scattered at my feet. I started sucking in the juices, but it was too late. There was no way to stop the embarrassment. I continued to answer questions about my career and the content of my lectures while everyone pretended that nothing was wrong. I had a strong desire to return home.

The department had a banquet in my honor during the first week of our visit with many members of the geology department and Mr. Zhou. We were introduced to our first "real" Chinese meal, which consisted of eel, duck (salt duck, a Nanjing specialty), sea slug, chicken, seaweed and peanuts. Dinner ended with dumplings, and everyone except me drank an evil-looking brown liquor. Beverly later told me that the liquor had a terrible taste. I told her I liked the sea slugs.

My first lecture started at 2 P.M. with Charlie as my interpreter. My main lesson was that the electrical power was not reliable. The power went off unexpectedly, and I gave my lecture without slides. I lectured every day for three weeks, six days a week, two hours a lecture, but it was no hardship, for I had come prepared with all my lectures typed. Unless the electricity shut down, the illustrations consisted of slides projected on the wall screen or on a white sheet that waved gently in the breeze. Charlie xeroxed a copy of my notes for every student, and they read each lecture before each class. Many of the students translated the words into Chinese characters directly onto the xeroxed pages. It was a successful teaching system because the lectures were reinforced by repetition. I lectured in English, Charlie interpreted my words into Chinese, and the students were able to read the lectures word for word, in Chinese and in English. My classes consisted of fifteen to twenty men and women. Most were young students from Nanjing, but some older professional geologists had come from other provinces.

After lectures we often entertained two or three Chinese professors in our apartment. They greatly enjoyed drinking our Coca-Colas which they could not buy with Chinese money because the Cokes were bottled in Hong Kong outside China. Chinese money could only be spent on Chinese products made in their country. "Coca-Cola is a great invention," Dr. Chen said smiling.

On some afternoons, Beverly and I walked through the city to look at the sights and people and record them with photographs. Many merchants sold their wares on the sidewalks. Some set up portable tables and benches to fix or make jewelry, mend shoes, or sell vegetables. Some sold boiled water in metal thermos bottles. There were hundreds of people walking on the sidewalks. They filled the stores, looking and buying. Before coming to China, I was under the impression that China, like Eastern European iron-curtain countries, had empty stores and people with no money, but Nanjing was sumptuous, crowded, and full of energy. I also expected the people to wear blue Mao Tse-tung uniforms. Far from it. In 1985, clothes were colorful and diverse because the central government had urged the Chinese citizens to support their rapidly growing garment industry. Many women wore modest jewelry. Men wore white shirts and slacks. On hot June days, men rolled up their pants to their knees, and some wore sandals without stockings. I saw one suave man with a Rolex watch.

Nanjing was a remarkably busy city. Buses and taxis traveled in the middle two lanes of the avenue flanked by bicycle lanes wide enough for ten bicycles abreast. The bicycle lanes were flanked by sidewalks full of people, many carrying baskets balanced across their shoulders on poles. In the streets were carters pulling loads of rock, sand, and coal. It was heartbreaking to watch them pull their heavy loads, leg muscles straining, the veins standing out in their necks and temples. On their bicycles, men and women carried cages of geese or ducks to market. Three-wheeled gasoline vehicles used as tractors in the fields moved at a swift twenty-mile-per-hour pace, hauling trailers full of vegetables as people on top held down the loads with their bodies and held their peaked hats on with their hands.

One morning before the lecture, we walked to the Friendship Store on Zhongshan Lu (Sun Yat-sen Boulevard) past Gulou (Drum Tower) Square, near a small park where old men sat and played cards surrounded by a menagerie of caged birds. We heard loud arguments. Charlie explained later: "The old men brag about the accomplishments of their birds while they play cards. Each one says his bird has the best song." Their throaty songs were hauntingly beautiful.

At one busy intersection, an immaculately dressed policeman directed automobile and bicycle traffic . His white gloves flashed in perfect rhythm as he gave signals to stop and permissions to turn, pass, or continue with exaggerated theatrical movements. He was a kinetic poem with a small audience of people watching him.

At the Friendship Store, only foreigners' Chinese currency was accepted, and we used some of it to buy an elegant natural pearl necklace for Beverly—price, the equivalent of less than eighty U.S. dollars. On our return, we walked through the narrow back streets, which still have the flavor of old Nanjing, with small dark houses and stacks of baskets and wooden planks. A small girl stared at us through a window yelling loudly for her mother, who whisked her away from the window and out of sight. In contrast, just a block away were recently built three- and four-story apartment buildings for the new age.

As we strolled along one of the main streets one Sunday, we saw a crowd of about one hundred people in a small park. A Chinese man approached us and in faultless English asked if we would like to meet some of his friends who practiced their English every Sunday at the "English Corner." The people had been motivated to learn English by Nixon's opening of relations between the United States and China. Listening to the Voice of America had aroused their curiosity about Americans and sharpened their English pronunciation. We agreed to join the English Corner and were immediately surrounded by the English-speaking Chinese enthusiasts who eagerly bombarded us with a smorgasbord of unrelated questions as Beverly and I stood back-to-back facing different audiences. Some of the questions to me were: "What is flextime?" "How much money do you make?" "What are your hobbies?" "What do you think of Richard Nixon?" "How far do you live from your work?" "How far is it from New York to Seattle?" "What do you eat for breakfast?" "What is a french fry?" To Beverly they asked, "How old are you?" "How many children do you have?" "Where do you work?" "Do you like your President Ronald Reagan?" "What do you like to eat?" "Do you drive a car? What is your hobby?" The questions were fascinating in themselves as they revealed an immense curiosity about American lives and idiosyncrasies. Little more than a decade before we arrived in China, the propaganda taught that Americans were foreign devils to be destroyed. But by 1985 we had become friends. We tired of the endless questions after about two hours, and they graciously released us from conversation with thank yous, goodbyes, and invitations to return.

On June 27, Dr. Chen announced, "The lectures are over and now we will go to see ignimbrites to the south." The destination was 250 miles south of Hangzhou (Hangchow) in country not yet open to foreign tourists. Dr. Chen had planned a very pleasant journey that included glimpses of Chinese culture. We traveled by train first to Wuxi, then to Shanghai, then to Hangzhou, and from there in a Toyota van to Yandang Mountain, our farthest point south, with several adventures on the way. With us were the driver of the van, two Ph.D. graduate students, Mr. Du and Mr. Pan, a geologist who had spent a year at the New Mexico School of Mines, Charlie, and Dr. Chen Kerong.

On our last night in Nanjing, Beverly and I invited our friends from the University to a banquet to thank them and the University for their kindness. Fourteen guests were there and the dinner menu, planned by them, was western—duck, chicken, mushi, steamed dumplings, squirrel fish, and for beverages, Haws wine and orange drink. Dessert was watermelon and ice cream. Total cost, eighty U.S. dollars.

Wuxi (woo-she), our first stop on the train trip, was a tourist stop. Our hotel overlooked a beautiful and serene lake, which we toured. But we wanted to see the Grand Canal, which had been started by Emperor Shih Huang Di two thousand years ago to unify north and south China and to avoid pirates on the sea. It began near Beijing, and parts of it still exist, with one segment in Wuxi. We argued that Wuxi is known as the Venice of China, but our hosts were reluctant to take us to the canal because it is in an old part of town without manicured tranquility. They finally relented and took us for a brief visit. It was a marvelous sight with the old boats carrying cargo, houses built in the water with smoke rising from some chimneys, and general robust activities of housewives and merchants—a visual feast of ancient China and very reminiscent of Venice, Italy, with foundations of buildings standing in the water.

For lunch, we went to Xishan Park where we had sweet lotus seed soup made with water from the second-best spring in all of China.

I asked, "Where is the first-best spring?"

Dr. Chen, Kerong said, "It doesn't matter because the first-best spring has dried up."

"Then isn't this the first-best spring?"

"No. It can only be the second-best spring because the place of the first-best spring is still there."

I changed the subject.

That afternoon, we left by train to Hangzhou, where we transferred our things into a Toyota van with a driver who took us on a wild ride. With constant honking, and without slowing for pedestrians in the small villages, we arrived at a single-story hotel similar to motels in the United States in Tonglu sixty miles south of Hangzhou. Dr. Chen, with considerable sensitivity, announced, "This is a place with a calm view of a beautiful river for Mrs. Fisher while she waits for us to come back from the field tomorrow." The river was magnificent, wide and picturesque, with fishermen standing in boats casting huge nets upon the calm, mirrored waters that reflected all the activity, doubling the visions. That evening, Beverly and I leaned over the railing of the hotel's patio to watch the fishermen pull their boats into the dock and sell fish to local residents.

It was Beverly's birthday, Monday, July 1. She spent a quiet day reading on the patio and gazing at the river while I took a field trip far into the countryside to look at volcanic rocks with the geologists. At one place I was astonished to see a bicyclist balancing a television in a basket over his back wheel, carrying it home. It seems as if the electrification of China by the Communists after the war brought a large part of China from medieval country life directly into the electronic age, bypassing the European evolution from the Industrial Revolution to the atomic age.

On the other hand, I clearly remember a stop at a quarry to look at granite that had forced its way through volcanic basement rocks. There were five men chiseling blasted granite boulders into large grinding wheels for animal-driven mills to grind grain for flour. Without special equipment except for small sledge hammers and hand chisels, they were making incredibly accurate mill wheels at least four feet in diameter and about ten inches thick with a square hole in the middle through which timbers were placed to turn them. I had previously seen stone mill wheels in European museums that had been used in the eighteenth century.

We returned to Shaoxing later that day and checked into a hotel that reminded us of the Oriental art museum in Pasadena, California. That evening, the Chinese geologists proudly entered our room with a birthday cake, joyfully and enthusiastically sang "Happy Birthday" to Beverly (even though her name was hard to pronounce), and then proceeded to help eat the cake with great gusto.

We later learned that Dr. Chen Kerong previously had traveled every mile of our trip south from Nanjing before our arrival in China. He had gotten permission from local authorities for us to travel through their districts and had made arrangements at hotels comfortable to westerners. Some of the officials exacted payment in one way or another, including requests for me to give a talk or an opinion about something.

From Shaoxing we went to Shangyu and stayed at the All County Government Hotel next to a canal and a railroad track where long strings of railway cars were being pulled by coal-burning steam engines. Women washed their clothes in the canal. We were visited on the first evening by a high local official who welcomed us to his beautiful province. We drank tea and talked as we sat on overstuffed couches.

The next day I was taken about forty miles south along a well-maintained gravel road through the countryside to a mining community. The purpose was to have me assess the operation of a meerschaum (clay) mine, and most important for them, I found out later, to assess the competence of their chief geologist. We arrived at a small village, the mine commune, and were met by the chief geologist, three subordinate geologists, and fifteen mine and state bureaucrats. Then we walked down a road lined on both sides with men, women, and children staring at me. I was the first westerner most of them had ever seen. Women held up their babies, pointed at me, explaining, I presumed, what I was. Some of the babies cried when they saw me. The workers had been given the day off in honor of my visit. We walked to a four-story building and climbed the stairs to the third floor to a large conference room with several large tables arranged in a rectangular pattern and eighteen chairs. Although it was a very hot day, at each place was the familiar thermos of hot boiled water. It was a long day.

First, the chief geologist, who spoke English, outlined the geology of the mine in tedious detail. He showed me maps of all the layers, many of which are volcanic, and described their compositions. His presentation took three hours and was punctuated by a few astute questions that he asked, most of which I could not answer. Like all good geologists, he readily admitted when he could not explain the features and many times asked for my opinion. Many men in his high position, in any country, would attempt to gloss over puzzling aspects of the geology and certainly would not admit to ignorance. He was refreshingly modest with an honest, skeptical scientific attitude. After lunch we spent two cold hours underground in the mine looking at the rocks and the mining operation.

At the end of the long day, I was given a completely unexpected surprise, a strong request equivalent to a demand, to evaluate the competence of the chief geologist. I was handed a notepad and pen and given a desk and chair to write my evaluation. With pleasure I gave an honest and glowing recommendation for the chief geologist, for he was truly excellent. I probably would have given a positive evaluation no matter what his competence, for I would not jeopardize a person's entire career on the basis of one day's evaluation.

On the long ride back from the mine, Charlie and I began talking about the United States and his experiences there. Suddenly he asked, "Why are there so many churches in the United States?" I answered that it might be because there are so many different kinds of Protestant Christians and they all want their own kind of church. Then he said, "What kind of Christian are you?"

I answered, "I'm not a Christian."

"Oh. A Buddhist?"

"No. I'm not a Buddhist, Muslim, Hindu, or Jew."

Charlie was puzzled. "Atheist?" he asked.

"Well, no. Not an agnostic either. I consider myself to be a spiritual humanist, for I am a product of the earth as are all living things." He looked as if he didn't quite believe me, for I am an American. I finished by saying, "I believe in the ideals of Jesus and Buddha about love and peace. Life has evolved out of the Universe, which can be thought of as God the mother and father."

"Wait until you see the Buddhist Temple at Yandang Mountain," he said.

With my duty to the mining company accomplished, we continued on our field trip south toward Yandang Mountain. I jotted some words in my notebook in order to recall the ride:

> Through beautiful countryside. Rice paddies. Small villages. Terraces lined by sweet potato vines. Hundreds of hilly acres of tea plants. Hillside mausoleum-like graves [we were told] of former lords and landowners. Canals. Winding mountain roads. Pottery huts with pottery ware displayed in front. Outdoor kilns in clay-rich countryside. Noodles drying on bridges and roadsides. Half-moon-shaped stone bridges. Hand-hewn ignimbrite-stone highway. Villages littered with motors and car parts. Lunch near an historical temple built in A.D. 598 from which Buddhism was exported to Japan. While waiting for road to clear after accident, photos of rice paddy workers and ignimbrite stone slab house. Yandang Mountain at 6:30 P.M.

We were in the heart of a craggy land carved from ignimbrite by the forces of the wind and ceaseless water erosion over eons of time. The hotel at which we stayed has been visited by thousands of Chinese tourists over the last one hundred years or more. We were in the Yandang Mountains (Wild Goose and Reed Marsh Mountains) in southeastern Zhejiang Province, about thirty miles from the coast of South China Sea.

Figure 23. Land carved from ignimbrite in Yandang Mountains, south-eastern China. Figure at bottom right is a farmer tending his sweet potato crop among the ancient spires of nature's spiritual offering. (Photo: Richard V. Fisher.)

A brochure in English read:

> Known for their natural beauty, the Yandang Mountains have some peaks rising straight from the ground into the clouds. Other peaks offer different images when looked at from different angles. The bizarre-shaped rocks are even more fascinating at twilight or under moonlight. The wonderful waterfalls tumble down their sides from sheer cliffs scores of meters high. The caves are either bright and spaceful palaced halls or twisted and intricating as if they were mysterious tunnels.

Charlie was right about Yandang Mountain. Earth and religion came together as a spiritual whole (Fig. 23). It was breathtaking. The erosional forms were like cathedrals, and one spire

Figure 24. Temple of worship within crevasse in ignimbrite for vertical support, Yandang Mountains, China. The five-story Buddhist temple is built within a fracture within the ignimbrite widened by erosion. A temple of man encompassed by nature's temple. (Photo: Richard V. Fisher.)

had a one hundred-foot cleft wide enough at the base to be filled with the first three levels of a large temple. The cleft narrowed upward and was filled with six levels of a temple (Fig. 24). We climbed the steps and, at the very top, arrived inside a cave made by the cleft. In the cave were golden icons of Buddha and Kuanyin (Guanyin), the Buddhist mother goddess. In the center, was a large bowl carved from the ignimbrite rock and filled by water that dripped from the ceiling. Next to the bowl was a porcelain cup. The water was holy and could be drunk from the bowl or from drips directly into the cup. It was the only place on our trip to China where the water did not need to be boiled.

When we first arrived at the hotel, we were in for a rude shock. Our room was very spacious and had magnificent views from all of the windows and the large balcony, but all of the metal screens were made of iron, rusted and torn with gaping holes. The mosquito netting over the bed was torn and covered by brown blotches of former squashed mosquitoes. The bathroom was large, made of gray concrete, with a toilet in one corner and a shower at one end that had no enclosure. I took a shower to discover what I should have noticed beforehand. The drain through which the water was supposed to flow was higher than the entire bathroom floor, and I had created a lake. To use the toilet, we had to walk through the lake. Luckily, the water was contained in the bathroom by a barrier at the entrance.

The day after we arrived, July 4, the geologists took me on a field trip to what they considered the source of the ignimbrite and the center of the caldera, but no vestige of a volcano shape had survived sixty million years of erosion. Escorted by military police in a jeep, we were taken to the top of Yandang Mountain 2,500 feet above sea level, to look at the textures and structures. The top of the mountain was shrouded in fog that swirled through the trees, adding a dash of mystery to the ancient mountain. The youngest rock of the region was a quartz-rich granite that filled a curved fracture forming part of a large ring pattern (a ring dike) of unknown diameter, thought to be the source of the ignimbrite of the surrounding countryside. The ignimbrite outside the ring was very thick and highly welded because of the original intense heat. We walked down a switchback trail through the silent swirling, ethereal fog, examining the layers as water dripped on us from the trees. In one place, at the base of a one hundred-foot-thick welded ignimbrite layer, was a sequence of very hard but thin wavy layers—deposits of a pyroclastic surge. The Chinese culture differs greatly from that of western cultures, but the ignimbrites are the same.

From the high ridge we could sometimes see through openings in the fog: rice paddies far below had terraces separated by sweet potato vines. Every square inch was cultivated. In places without rice paddies, watermelons were growing. With that discovery, I realized why men climbed up the steep one thousand-foot slopes to cut ferns and carry them on their backs in big bundles down to the fields. They used the ferns to pamper the watermelons. They made soft padding beneath the melons so that they would not touch the ground to distort their shape or color. Periodically they would rotate the melons so they would ripen evenly.

We walked two or three miles back to the van. On the way, we encountered a small village of several houses with thatched roofs on two sides of a wide trail too small for vehicular traffic. Again I became a visiting celebrity. Somehow, while we were walking toward the village, the residents had heard about the foreigner who was going to pass through. Everyone living in the village lined the pathway to see me, including the men who came in from the fields. I kept nodding and saying hello in Chinese. "Ni hao. Ni hao. Ni hao. Ni hao." Over and over. As if it were a celebration, I was greeted with smiles and laughter at my attempts to say hello in Chinese, while small children averted their eyes.

That evening we were visited by the local governor of the district. The government official and his assistant, the hotel manager, Charlie and Dr. Chen, myself, and Beverly sat in a room with overstuffed chairs and a sofa, which was part of our hotel "suite." We drank hot tea and told the official how much we liked China. He then got around to his business—checking up on the hotel manager—and he asked, "Tell me what you think of the hotel?"

I was not going to tell him that it was shabby, run down, and poorly built, causing the manager great embarrassment and perhaps resulting in imprisonment. I meant him no harm. Surrounded by unequaled beauty, we could ignore the inconveniences.

I said truthfully, "The hotel is located in the most beautiful place that I have ever been. More beautiful than anywhere in the United States, Japan, Europe, or other countries that I have visited."

The governor was pleased and didn't seem to mind that I didn't answer his question. I must confess that in forty-four days

I learned more about the amazing Chinese people than I did about volcanology, but that was essentially the reason I had accepted the challenge to lecture there. They are energetic and highly intelligent people, and we found them curious and friendly—no different in their hopes for a good life and love of family than the average person in the United States.

One commentary on the food that we ate in mainland China was not realized until we returned to westernized Hong Kong: The first night back, we went to a restaurant to have a familiar western meal. Although we did not know it, our bodies craved fresh vegetables. We ordered nothing more for supper than large dishes of fresh salad to celebrate our return to the Western world.

Chapter Fifteen

Italy and Ignimbrite

Italy, 1984 and 1988

I wasn't prepared for volcanological fieldwork in a large urban area, but that's where the volcano is—Pozzuoli and Naples, Italy, are in the crater. One and a half million people live packed together within a large caldera recognizable only by volcanologists after intensive tracing out of formations and fractures. Mount Vesuvius is dwarfed by the caldera. Cliffs that border Naples and Pozzuoli to the north are part of the crater rim. The southern rim is beneath the waters of the Bay of Naples and the Pozzuoli embayment. In addition to people living eye-to-eye, there are several small volcanoes within the region of the crater. Some of my work had to be done in Naples and Pozzuoli, but thankfully, most of the study was outside the city in the farmlands and foothills of the Apennine Mountains on an ignimbrite layer that once totally blanketed the countryside. The ignimbrite came from the Pozzuoli-Naples caldera.

About thirty-seven thousand years ago, there was a gigantic explosive eruption, probably centered in what is now called Pozzuoli Bay. The eruption produced a pyroclastic flow of such great volume (120 cubic miles), plus an unknown quantity of ash that spread across the Mediterranean Sea, that it caused the ground to collapse and created a caldera. The pyroclastic flow traveled outward from the explosion center over the Tyrrhenian Sea, the Bay of Naples, along the coast north of Pozzuoli-Naples, and eastward into the Apennine Mountains, leaving behind a vast sheet of ignimbrite (the Campanian ignimbrite) destroying all life wherever it went. The ignimbrite filled valleys in the foothills of the mountains, filled the crater of Roccamonfina Volcano some thirty-five miles north of Pozzuoli Bay, and filled in the flat, fertile plain now drained by the Volturno River. It extended thirty-five miles east to Salerno, across the land on which Vesuvius Volcano was later built, and it flowed across the waters of the Bay of Naples to blanket the Sorrento Peninsula. The vacation town of Sorrento is built on Campanian ignimbrite that is over one hundred feet thick.

When I first visited southern Italy in January 1981 while studying in Germany, I examined the Campanian ignimbrite in a few places around Naples and Sorrento. I was fascinated by its wide distribution and wondered how the fragments in the pyroclastic flow moved over the mountains to deposit the ignimbrite, and moved over the waters of the Bay of Naples onto the Sorrento Peninsula. In 1984, I revisited the region to investigate the feasibility of future research. In 1988, I returned to Italy to begin research financed by a grant from the National Geographic Society, awarded to Dr. Grant Heiken and myself. I later collaborated with Drs. Giovanni Orsi and Lucia Civetta (from the University of Naples), Heiken, and Dr. Michael Ort with research funds from the National Science Foundation awarded in 1990.

The volcano is still alive. It has a long life span compared to human beings. Near the center of the caldera, in the town of Pozzuoli (Fig. 25), repeated earthquakes and ground-swelling episodes have occured since the time of the Romans and no doubt before that. The latest period of seismic activity and swelling was from 1982 to 1985, when the activity subsided to a low level.

Pozzuoli was in a state of crisis when I visited there in 1984. Between June 1982 and December 1984, the magma body beneath had raised the center of Pozzuoli six feet. There were cracks in the buildings everywhere I went in the city streets, which were not nearly as deserted as I thought they should be.

Figure 25. Pozzuoli, Italy, situated in the Campanian caldera. This is a view of the city of Pozzuoli, which is at or near the center of the caldera that erupted the Campanian ignimbrite thirty-seven thousand years ago. The photograph was taken in April 1984, when the volcano's center was slowly rising. As the center rose, buildings cracked, plaster fell, doors and windows would not shut, windows broke, and people moved out of the city or lived in tents and house trailers. (Photo: Richard V. Fisher.)

As the land rose inch by inch, the buildings were pushed out of line so that doors would not shut and windows would not close (Fig. 26). Some walls had fallen, and many people had vacated their homes. Stopping at a small coffee bar that was still open, I met a man named Antonio who could speak English, and I asked him, "Do you live in the city?"

"All my life I have lived in Pozzuoli, and now the government says I must sell my apartment. Now I live in a trailer in the park along the waterfront." He pointed at the temporary cluster of house trailers where children played and wives hung the clothes from lines that crossed from trailer to trailer (Fig. 27). "The authorities say there might be an eruption but it is nonsense. The

Figure 26. Pozzuoli, Italy. Damaged buildings were everywhere throughout Pozzuoli in April 1984. Some were merely cracked, but others had major damage, and some were supported temporarily by buttresses. There was traffic, though far less than normal. (Photo: Richard V. Fisher.)

government wants me to believe that, so they can buy property at half price."

"But haven't you felt the earthquakes? And isn't your building slowly cracking and falling?" I asked.

"Yes, but they want me to buy a new apartment made of ugly concrete across the hills on barren land. I am a fisherman and I need to be by my sea and live in my place of beauty. I will not move from Pozzuoli!"

I could understand his frustration, but his total denial of possible annihilation was based on groundless logic. He wanted to blame his predicament on the government. If he could assume that the authorities were corrupt, then he could believe that they were only pretending there was going to be an eruption. If that

159

Figure 27. Tent and trailer city, Pozzuoli, Italy. Open parkland along Pozzuoli's waterfront was used for temporary housing in tents and trailers, April 1984. Damaged buildings are in the background.
(Photo: Richard V. Fisher.)

were so, there would be no eruption and, therefore, no reason to move. That kind of logic explains why it is so difficult to evacuate people in time of crises.

Another certain sign that the volcano is alive is that, within the crater of Solfatara volcano, not far from Pozzuoli, is an area of hot springs, boiling mud pits, and sulfurous steam vents caused by heat that rises from the cooling magma body. The heat comes in contact with groundwater derived from rainwater that seeps into the rocks. Solfatara crater is one of the small volcanoes that occupy the caldera.

I started fieldwork in 1988. This consisted of driving through heavy traffic in the cities to look for new excavations along streets or building construction sites where exposures were fresh—and available if permission could be obtained—to examine and describe the rocks. Most work was, fortunately, on farmland and hillsides north, east, and south of Naples.

I enjoyed working with the Italians. No matter if we were in the big city or in the countryside, at about ten o'clock, my partner of the day would say "Coffee?" In the rural areas, there was always a small village nearby with a coffee bar. Espresso is the coffee drink of choice, and for the first coffee break, I expected to sit and sip coffee with conversation to review the morning's work. But Italians do not sip their espresso, they drink it like a shot of whiskey, then it's "Let's go!"

Best of all was lunch. In every city and every village in rural areas are small markets run by momma and poppa. On the very first day that I began fieldwork in Italy, Claudio, my field assistant, suggested that we have a paninno. I watched Claudio order his, then I ordered. I stood in front of the glass case that held the sandwich items, and I pointed at what I wanted between the two halves of fresh Italian bread I had chosen. First was the Bufala mozarella cheese, which is food made in heaven—absolutely the best food I have ever tasted anywhere in the world. It is made from water-buffalo milk only from the Naples region. Then, I could choose tomatoes, picked that morning from a nearby farm. Sometimes I would include prosciutto or mortadella, a type of minced meat. Then we would find a spot along a street or road, and if in the mountains, a place with a view, to sit on the ground and eat lunch.

The distribution of the Campanian ignimbrite is a fascinating puzzle because it traveled sixty miles or more in all directions and surmounted three thousand-foot mountains. Despite extensive erosion over the years, there are plenty of outcrops, although most are on private property because of the dense population. Because I neither speak nor understand Italian, an Italian-speaking field assistant was necessary.

Armed with the idea that there appear to be two ways that pyroclastic currents move across the landscape—as expanded flows, thicker than the height of the mountains they cross, or as dense, nonexpanded, nonturbulent ground-hugging sheets—I went to Italy to find out how far and where Campanian ignimbrite went and to determine its flow directions.

When Beverly and I arrived in Italy in May 1988, Dr. Giovanni Orsi had arranged for us to rent a house in the village of Praiano along the Amalfi coast. The place was paradise. The village is built on a steep limestone mountainside that rises two thousand feet out of the sea. The house was surrounded by a garden and a small grove of orange trees. A balcony ran the length of the house, and the front room and both bedrooms had access to the balcony. To the east and west, we could see gardens, a few red-tiled rooftops, and in the distance, a ridge with houses terraced to the top of the mountain. Each room, including the kitchen, overlooked the Tyrrhenian Sea with the horizon far away, stretching from the rocky coast six hundred to seven hundred feet down to the water. On nights of the full moon, we could watch it cross the sky from our bedroom.

The entire Amalfi coast is eroded into limestone cliffs, and all of the villages along the coast, which have been here for centuries, are terraced down the slopes to the sea. The coast and its villages are a collage of architectural and landscape art that has evolved over two thousand years. No single mortal could have designed it.

There are only three drawbacks to paradise. (1) The entire Amalfi coast is connected by a narrow paved road that follows every curve and contour that God ever invented, and it is host to huge tour buses from all over Europe that can barely pass one another, cannot turn around, and cannot negotiate some tight turns in the road without going forward and backward a few times. Often there is gridlock when two tour buses meet in one of the many tunnels cut through projecting ridges along the rocky coast. In addition, every young Italian in Italy thinks he is Mario Andretti racing in the Grand Prix. (2) It is very hot and humid in the summer, especially in August. (3) It is fairly crowded in May when tourist season begins, but there are ten times as many tourists in July and August.

At first I thought I could work by myself in the region, and I had one of the Italian professors write me a note on official university paper explaining in Italian that I was on official business and that I needed to look at rocks on their property. Two days after getting the note, I needed to examine some rock outcrops on a small farm at the foot of Mount Vesuvius. An old Italian woman began to accost me in agitated Italian when she saw me on her property. Trying to look official, I gave her the note explaining why I was there. She took the paper, squinted at it for a minute and then turned and began to yell toward her house. A young girl appeared, clearly embarrassed, and walked slowly toward us as the old woman continued talking loudly and rapidly. The young girl took the note and explained what I had already suspected. "My grandmother cannot read," she said softly. "I learn English in school. Grandmother never went to school."

A Ph.D. student of Giovanni's, Claudio, became my field assistant. We spent four months driving through the greater Naples region searching for basal contacts of the ignimbrite on older rocks of the region. During those four months, we traveled ten thousand miles within the foothills of the Appenines and discovered fourteen good basal sections of the layer.

I pursued the basal sections of ignimbrite because the rocks on the bottom "talked" to me. Throughout most of a thick ignimbrite layer, there are few physical features in the rocks that give clues to flow mechanisms—usually, the visible features in the rocks are monotonously the same. There is no obvious layering, and there are very few changes in fragment size from the top to near the base. Physical evidence of flow usually occurs at the base. Pyroclastic flows moving at hurricane velocities scour the surface, and fragments from the ground can become mixed within the lower part of the layer. If the ground over which they move, for example, is composed of limestone, the lower part of the ignimbrite will contain bits and pieces of the limestone.

The presence of a nonmoving boundary, such as the ground surface, affects the overlying flow. The shearing motion of the flow over the surface may cause turbulence to develop in the flow next to the ground. Turbulence causes the flow to expand, and particles that fall out move and bounce along the surface in what has been called "traction flow." This in turn causes thin layers to spread out across the ground in the direction of flow. It also can cause the beds to be wavy in shape and to develop cross bedding.

If a pyroclastic flow moves slowly and in a ponderous way, like concrete flowing down the chute of a concrete truck, mixing of surface particles into the moving flow will rarely occur unless there are particles light enough to float. In addition, thin bedding or cross bedding will not develop from such a fluid. This does not tell how fast the pyroclastic flow moved but allows me to speculate that movement was turbulent or nonturbulent (dilute or dense).

One day when my assistant was sick, I went to a rock quarry east of Naples. It was very busy with trucks arriving empty and others leaving loaded with crushed rock. Dodging one truck after another, I found the headquarters building and the quarry manager. At the back of the quarry, I could see an outcrop that looked like the Campanian ignimbrite, but I needed to examine the rock from close range to confirm it. He could not understand English, but an office worker acted as interpreter.

"May I examine the rocks behind the parked trucks?" I said pointing at them through the window.

With a frown he asked, "Why?"

"There was a big eruption of a volcano at Pozzuoli many thousands of years ago, and I am trying to find out how far the ash traveled. I am doing research for the University of Naples."

He was puzzled and suspicious. Rocks carried minerals and could be valuable, Again he said, "Why? Why do you do research on those rocks?"

I said, "I am trying to find out how big the eruption was."

And again he asked, "Why? Why do you want to know that?"

"So that we can predict where it might go next time." I explained that it came across the land like a hurricane. "It's for the safety of the people. If I can tell how the flow moved, I can tell people where to go to get out the way of danger if another eruption begins."

"But why do you want to look at the rocks over there." He could not understand why anyone would want to look at a rock outcrop unless there was something valuable about the rock. He thought that I had a hidden motive.

"The rocks give information about how the volcano hurricane moved and how far it went."

"But they are just rocks. And how could they come from another place? They have always been there."

I realized that if he believed that, then none of my reasons could make any sense to him. Moreover, as a practical man, he thought that rocks were only important economically; therefore I must not be telling the truth. He became increasingly agitated because I would not tell him the real reason I wanted to look at the rocks. I was wasting his time. "I am a busy man," he said. "You cannot look at the rocks."

One day Claudio and I were looking for basal contacts of the Campanian ignimbrite along the Titerno River near the village of Faicchio, thirty miles northeast of Naples. Searching for routes to the river, we came to a farm that bordered the river, but we had to walk across the property to reach it. Claudio got permission for us to cross, and along the low cliffs of the river, we discovered a half-mile-long exposure that showed the basal contact perfectly. We spent the morning describing the rocks and then headed back at about two o'clock for the three-hour drive back home.

As we walked in front of the farmhouse to the car, the farmer invited us to have lunch with his family. At first we declined, but the farmer was persistent.

We were in for a two-hour treat! The grandmother lived with the farmer, his wife, and three children, one of whom was in high school and wanted to practice his English. The farmer's sister and his brother-in-law were also there on vacation from Milano. It was a festive occasion in the middle of the week. The walls of the house were covered with old photographs of relatives and religious prints with old and ornate frames. Icons of Jesus and Mary decorated the mantle over a huge fireplace. Old worn-out rugs were on the floor. A simple light fixture hung from the ceiling. The musty old house smelled of sour wine, pasta, bread, garlic, and onions, a very pleasant and homey odor.

The table was big and it was full of food grown on the farm. The red wine was made from their grapes. There was a huge tomato and lettuce salad from their garden. Pasta with red tomato sauce and cheese in large bowls was the first course, and the main course was roasted goat meat. Dessert was fruit from their trees.

It was a noisy meal with everyone talking. The boy said to me, "Hello. How are you?" I said, "I am fine. How are you?" He said, "I am fine, thank you. What is your name?" "Richard," I answered. "What is yours?" "Mario," he said.

Grandmother Petrini sat beside Mario, smiling with great pride. As translated by Claudio, she said, "I am so proud of Mario. He speaks English."

At the same time, father Petrini and his brother were having a loud argument, gesturing wildly in the Italian way. They both looked angry, and I thought that they would come to blows. "Why are they fighting?" I asked Claudio.

"Oh, no," he said, "they're not fighting. They are just discussing when to plant the tomatoes."

From that day on, while in Italy, I remained unconcerned about the many "angry arguments" so common between Italians.

Italy, 1990

My study of the Campanian ignimbrite expanded when Drs. Giovanni Orsi and Lucia Civetta, my friend and associate Dr. Grant Heiken, and I wrote research grants to study the emplacement processes of the ignimbrite, our main interest, and the chemistry of the layer to identify subterranean chemical processes within the magma. Dr. Michael Ort joined later to study magnetic susceptibility aspects of the layer to determine flow directions, as he had done with ignimbrite in the Andes. The weak magnetic signature of the rocks, as determined by this method of measurement, can indicate the direction of movement of flow and has turned out to be an important tool in interpreting emplacement processes.

With new funding, and using our own money for side trips and Beverly's expenses, Beverly and I returned to Italy in 1990 so that I could complete the research started in 1988. We arrived in Rome June 1, rented a car, and drove south to meet Giovanni Orsi in a small village on the Sorrento Peninsula. He took us to an apartment he had reserved for us near the center of the town of Sorrento, where we were to stay for the summer.

We immediately became immersed in and surrounded by traditional Italy. Sorrento is a famous and ancient vacation city, now too crowded with tourists, cars, and souvenir shops. The churches are full of ancient history and are havens of quiet from the noise of the streets. The gardens are beautiful, as are the ports down by the sea where there are ships to Naples, Capri, and Ischia. There are trains to Pompeii, Herculaneum, Rome, Florence, almost anywhere.

The apartment was in a large complex on the Avenue of the Oranges, via degli Arancia, almost in the heart of downtown. At night we would stroll through the streets, visit the ice cream parlor, or lean on the rail of the terrace that overlooks the harbor and across the Bay of Naples. On clear nights, the lights of Naples shone brightly twenty miles away, and Mount Vesuvius was visible at the eastern end of the bay.

The apartment had several rooms, more than we needed. We think that a couple of old aunts of the family moved out so that they could earn our rent for two months. It had beautiful Italian-tiled floors, some tiled walls, and many windows overlooking the orange, lemon, and olive groves that marched up the mountain behind us, and balconies in front that looked down on the busy Street of Oranges. We slept in an antique bed that sagged in all the wrong places but was ornately carved. We were surrounded by madonnas, wax angels, santos, and christos, and some very old relatives watching us from their silver and pearl frames on the dresser—our Italian family for two months.

During the summer of 1990, I worked alone or, at times, with Giovanni. We discovered more basal outcrops and finished the network of stations needed to understand the emplacement of the ignimbrite. We found that the Campanian pyroclastic flow traveled over ridges higher than three thousand feet and filled valleys on the other sides. The magnetic information showed that the Campanian flow everywhere had moved down rivers, valleys, and all slopes. This means that the volcanic debris comprising the ignimbrite was brought to the area of deposition by the expanded turbulent transporting cloud from the Pozzuoli region, and when large volumes of particles fell to the ground, they drained downhill toward the lowlands.

The best evidence that the Campanian pyroclastic transport system was an expanded flow is the fact that its deposit, the Campanian ignimbrite, is more than 120 feet thick on the north side of the Sorrento Peninsula, some twenty miles over the water from Pozzuoli Bay, the center of the eruption. The flow must have been expanded because it traveled over water. Only particles at the base of the flow sank as they touched the water surface.

A similar event occurred at the Krakatau, Java, eruption of 1883. During that eruption, the upper part of the Krakatau pyroclastic flow remained less dense than seawater and traveled more than thirty miles over open water to the Sumatra coast, where about two thousand people died of burns from the flow.

Conclusions drawn from the flow behavior of the Campanian ignimbrite complemented my studies of pyroclastic flows and surges on the John Day Formation, Oregon, Mount Pelée, Martinique, Laacher See, Germany and Mount St. Helens, as well as research on base surge deposits in Germany, Hawaii, New Zealand, Mexico, and the western United States. I have made a beginning toward solving a complex problem, but an enormous amount of research must be done before a satisfactory understanding can be reached. Time will tell whether or not I was on the right track.

Suggested Reading

Furneaux, R., *Krakatoa*. Englewood Cliffs, N.J.: Prentice-Hall, 1964.

Epilogue

Time passes slowly while one lives in it, but retrospective time flies at the speed of thought. My thirty-eight years of teaching and research went in the blink of an eye. One day in July 1955, I stirred up a cloud of dust as I drove across an empty field (now occupied by Engineering, Physics, Geology, and Chemistry buildings) to the front entrance of the new science building at Santa Barbara College (now University of California, Santa Barbara), parked beneath the windows at the front entrance, and joined the geology faculty in person. The campus had 1,200 students at that time and consisted largely of barracks, which had housed Marine Air Corps pilots during World War II. I was twenty-six years old.

When I awoke the next day, it was January 1, 1993, and I had retired. The years 1955 to 1993 had become a kaleidoscope of interconnected people and events. In this book I have connected my research projects into a series of (sometimes obscurely) related events. One project led the way to another, a characteristic of advancing scientific research.

During my early years at the university, I was most attracted to creative research and writing the results of my research for the geological profession. It was equally gratifying to pass research results on to beginning students as well as graduate students and involve them in creative thinking. In my later years, through writing, I have become interested in teaching larger audiences about volcanoes and the importance of supporting research in volcanology.

Lacking formal classroom teaching assignments, I turned to popular writing as a teaching device. I collaborated with Grant Heiken of Los Alamos National Laboratory, and Jeffrey Hulen of The Energy and Geoscience Institute at the University of Utah, to write a book for laypeople about the interaction of volcanoes and people (*Volcanoes: Crucibles of Change*, Princeton University Press, 1997). The present book, *Out of the Crater: Chronicles of a Volcanologist*, also reaches out to the public. I am therefore able to teach about volcanoes to a potentially far larger audience than I could find in a single classroom. This is also accomplished through my World Wide Website, the "Volcano Information Center" (http://www.geol.ucsb.edu/~fisher). Judging from a few E-mail responses to the web page, this website is used by nongeologists interested in learning about volcanoes and by students from the third grade to college level who write class reports on various aspects of volcanology.

I did not quit research activities upon retirement. I am collaborating with Dr. Jean-Luc Schneider, who teaches at Lille University, France, on a project in the Auvergne District. Under Jean-Luc's direction, we are studying the flow mechanisms of collapse debris avalanche deposits at Cantal volcano. Jean-Luc spent a year as a postdoctoral candidate with me at Santa Barbara in 1991. Another ongoing collaboration is with Drs. Giovanni Orsi and Lucia Civetta (professors at the University of Naples) on a project concerned with the volcanic hazards of the Phlegrean Fields caldera in the region of Naples, Italy. Dr. Civetta is the director of the Vesuvius Volcano Observatory.

My research has approached the problem of mass flows from field observations of the features in rock layers that give evidence of the movement and emplacement of particles; what were the influencing processes of flow and emplacement that caused individual particles to come together in their final resting pattern? I study old deposits of volcanic eruptions to determine how they were formed, while other volcanologists, with more gripping stories to tell, do their research in craters and on the slopes of erupting volcanoes. The rock outcrops have spoken to me, and they have said about everything possible in my realm of understanding. Now, with some regret and envy as I step aside, I watch younger men and women wrestle with some of the same puzzling problems that I tackled during my research years. I hope that the younger scientists remember the contributions of those who have paved the way for them. Although it is necessary to become aquainted with the rocks in the field, modern research concentrates more and more on computer simulations and experimental studies of small-scale sediment gravity flows in laboratory flumes to help solve further problems. I look forward with great anticipation to reading the answers to some of the questions that I have pursued during my lifetime.

The unexpected surprise of my career is how seemingly unrelated events wove themselves into a sequence of interrelated research studies along a sinuous pathway with many twists, turns, and blind alleys. My path in science started with reading dinosaur books in the 1930s, exploring the Puente Hills, learning about science from Kenneth Gobalet in the seventh grade, and witnessing the first base surge at Bikini in 1946, before starting college and being introduced to geology. Events and experiences have continuously folded back upon themselves to influence the path's directions. Whatever I had done or thought or witnessed or learned in the past was brought to bear on current problems, and the current problems influenced future studies.

Many of us wonder what influenced us to follow our particular professional path. For me, the main driving force has been an incessant and persistent voice asking, "How?" "Why?" I recognize this voice as that of my grandmother Miller. In 1935, when I was teased for asking "why" so much, she held me and validated my curiosity by saying that she was proud that I could ask why, and when I was ten, she reinforced it with this short poem: "Richard. Waving his arms pointing to the sky. Always asking why?"

Glossary

aa lava: Lava of basaltic composition that has a very rough surface (contrast with pahoehoe lava)

ash: Fine granular pyroclastic material, with grains 0.04 inches (2.0 mm) or less in diameter. Hardened ash is called *tuff*.

basalt: Lava, usually black or very dark gray, that has 55 percent silica or less. Ocean floors of the world are covered by basalt lava flows.

base surge: A ring-shaped cloud of gas and suspended solid debris that moves (surges) radially outward from the base of a vertical explosion column at high velocity as a density flow.

blast surge: A blast or explosion directed outward from a volcano with great force.

block: A solid volcanic rock (pyroclast) ejected from a volcano and having a diameter greater than 2.6 inches (64 mm).

bomb: A viscous particle (pyroclast), of a diameter is greater than 2.6 inches (64 mm), ejected from a volcano before the particle is frozen solid; shapes vary greatly, including spindles, cow pies, ribbons, etc.

caldera: A crater larger than five miles in diameter formed by collapse of the ground or volcano into an underground cavity formed by the ejection of an equivalent amount of pyroclastic material. Some calderas are larger than forty miles in diameter.

cinder: A rough-surfaced pyroclastic fragment baseball to nut size, formed directly from magma and explosively ejected to the surface. It is gaseous and therefore many bubbles (vesicles) form within the viscous material; synonomous with scoria.

cinder cone: A volcano formed by cinders that pile up around an active vent.

cross beds: A group of layered laminations (very thin layers) truncated by other layers across eroded edges within a larger layer, or within an assemblage of layers.

dome: A dome-shaped protrusion of solidified magma formed by slow extrusion of highly viscous magma located on the side of, or within, the crater of a volcano.

ejecta: Pyroclastic material explosively ejected from a volcano.

fallout: Accumulation of pyroclastic material that falls from the sky.

fumarole: A vent from which gases from magma—or steam created by hot magma heating water—escape into the atmosphere.

hydrovolcanic eruption: An explosion or eruption caused by sudden expansion of water when mixed with magma.

ignimbrite: A deposit of a pyroclastic flow or pyroclastic surge or both.

lapilli: Pyroclastic particles ranging in size from 0.04 to 2.6 inches (2 to 64 mm).

laterite: Soil so rich in iron oxide that it may be mined as an iron ore.

lava: A volcanic effusion of molten rock that flows upon the ground; also solidified lava.

limestone: A rock composed of an aggregate of calcium carbonate minerals.

littoral cone: A mound of volcanic particles formed when hot lava mixes with water at the shoreline of a body of water and explodes; the mound is a cone that looks like a volcano but has no underground vent.

maar: A small volcano with wide diameter relative to its height, often with a bowl-shaped crater.

magma: Molten rock with contained gases that forms within or somewhat below the earth's crust. It is ejected in highly explosive eruptions to form pyroclasts or lava flows.

magmatic eruption: An explosive eruptions from the sudden expansion of magma from internal gas pressure.

nuée ardente: A hot, heavier-than-air, turbulent "volcanic hurricane" composed of a mixture of hot gases and solid particles. Nuées ardentes flow rapidly down the slopes of a volcano often attaining speeds fifty to one hundred miles an hour. Such flows commonly form in two parts, a lower "glowing avalanche," and an upper "glowing cloud."

pahoehoe lava: Lava that has a smooth surface after cooling, compared with the roughness of aa lava.

palagonite: A yellow waxy mineraloid composed of clay and iron oxides, formed by the weathering and breakdown of basalt glass.

Peléean eruption: An eruption that produces a nuée ardente, such as that at Mount Pelée in 1902.

Plinian eruption: An eruption that forms ash clouds above the volcano to heights as much as thirty miles into the atmosphere.

pumice: A glass foam formed by frothing gas-rich magma. Pumice is very lightweight and can float in water.

pyroclastic: Refers to the fragmental materials formed by explosive processes of volcanoes.

pyroclastic flow: A heavier-than-air, hurricanelike, nonturbulent cloud of volcanic particles mixed with hot gases; more dense than a pyroclastic surge.

pyroclastic surge: A heavier-than-air, hurricanelike, turbulent cloud of volcanic particles mixed with hot gases; less dense than a pyroclastic flow.

sandstone: A sedimetary rock formed of cemented grains with diameters less than 0.04 inch (2 mm).

sedimentary: Refers to various aspects of sediments, which are layered deposits of rock and mineral particles.

scoria: A rough-surfaced pyroclastic fragment, baseball to nut size, formed directly from magma and explosively ejected to the surface. It is gaseous and therefore many bubbles (vesicles) form within the viscous material; synonymous with cinder.

seismograph: An instrument that records earthquakes.

spatter cone: A small cone made of hot, gas-rich bombs of fluid basalt lava that fall and stick together around a vent.

strata: Layers of rock are said to be stratified and are referred to in the plural as strata; an individual layer is a stratum.

Strombolian eruption: An eruption that produces high-arcing, incandescent "rooster-tails" reminiscent of colorful fireworks fountains. The ejecta are mainly basaltic cinders and bombs that construct cinder cones.

tsunami: Large waves formed by large undersea landslides commonly caused by earthquakes. From the Japanese word for seismic sea wave (misnamed "tidal wave," although it is not related to tides).

tuff: Volcanic ash hardened by pressure or from natural cements deposited from water that seeps through the ground.

volcanic: Pertains to a volcano, e.g., volcanic rock, volcanic eruption.

volcaniclastic: Refers to clastic rocks composed of volcanic particles.

volcanic ash: Fragmental pieces of glass, crystals, and rock fragments, less than 0.4 inch (2mm) in diameter, extruded by volcanoes.

volcanic sandstone: A rock composed mostly of sand-sized volcanic particles.

Index

Alcaraz, Arturo (Taal, Philippines), 50
Alto Plano (Siberia Argentina, high
    Andes), 128; Cusi Cusi village, 130
Anasazi (native Americans, N.M.), 9
Andal, Contrado (Taal volcano), 57
Armstrong, Neil, landing on the moon,
    48–49
Auvergne volcanic district (France), 86;
    Puy de Dome, 86; Cantal volcano, 171
Azores: volcanological meeting, 85; Lom-
    bosada Water Works, San Miguel Is-
    land, 85

Bikini, atom bomb tests, 15; Able Day, det-
    onation above water, 16–17; atoll, 16;
    Baker Day, detonation under water, 18–
    19; first base surge, 19–21; target ships
    of atom bombs, 16, 17–18
breakfast in Japan, 56

caldera, 11, 126; Cerro Panizos, 128–134;
    Long Valley (Calif.), 126–127; Los
    Alamos (N.M.), 126; Phlegrean Fields
    (Naples), 127, 171; Pozzuoli-Naples cal-
    dera, 156; uplift of central Pozzuoli by
    magma, 157–160; Yellowstone National
    Park (Wyo.) 127–128
Campanian ignimbrite: Claudio, field as-
    sistant, 1988, 163; distribution of, 161;
    eruption from Pozzuoli Bay, 156–157;
    transport system, evidence of ex-
    panded flow, 168
Carey, Steven, 111, 112, 113, 115
Cascade Mountains, author's Ph.D. dis-
    sertation, 25–26
Cerro Panizos caldera (high Andes), 128–
    129; Abutarda Lake, exotic birds, 131–
    132; field trip to, 129–130; Michael
    Ort's thesis, 128; Puesto Zenon Coria,
    stone hut, final destination, 133–134;
    trip back to Cusi Cusi, 134
China: accommodations in Nanjing, 141;

country village, 154; evaluation of chief
    geologist, Shangyu mine, 150; food in,
    142; Hangzhou, 147; 154; impressions
    of Shanghai, 138; lectures at Nanjing
    University, 142–143; return to Hong
    Kong, 155; Shangyu and trip to mine,
    148–149; Shaoxing, Beverly's birthday
    party, 148; sights of Nanjing, 143–145;
    Tonglu, 147; train to Nanjing, 139–140;
    Wuxi, 146; Yandang ignimbrite, 153;
    Yandang Mountain and resort, 150–153
Civetta, Lucia (Vesuvius Volcanological
    Observatory), 157, 171
Cornell, Winton, 111
Crowe, Bruce, 61

Datuin, Roger (Taal, Philippines), 58

Eifel volcanic district (Germany), 81

El Chichón volcano (Mexico): adventur-
    ous return from, 121–125; air flight to,
    116; chance meeting with Maurice and
    Katia Krafft, 118–119; crater lake, 120–
    121; destruction of Vulcan Chichónal
    village, 117–118; eruption of, 119–121;
    field of pyroclastic surge dunes, 116; Pi-
    chucalco, 115
Europe, volcanic districts in, 84–85

Fort de France (Martinique), 62
Fort MacArthur Induction Center (Calif.),
    13

Gaia, 4, 5
Gardello, David, 115
Geiger counter, 15
geologic time, 3
Gobalet, Kenneth (seventh grade science
    teacher, 1940), 23–24

Hans-Ulrich Schmincke (Bochum, Ger-
    many), 80

ABOUT THE AUTHOR

Richard V. Fisher is Professor Emeritus of Geological Sciences at the University of California, Santa Barbara, where he has taught and researched since 1955. In 1997, he was awarded the Thorarinsson Medal, the highest honor of the International Society of Volcanologists. Fisher is coauthor, with Grant Heiken and Jeffrey B. Hulen, of *Volcanoes: Crucibles of Change* (Princeton). He also wrote *Pyroclastic Rocks* with H.-U. Schmincke and coedited *Sedimentation in Volcanic Settings* with G. A. Smith.